TIPS ON BUYING CESSNA SINGLES

by

Alton K. Marsh, CFII

Cessna Skyhawk

PUBLISHED BY AVIATION BOOK CO.

NOTE

PRICES QUOTED IN THIS BOOK WERE APPROXIMATELY CORRECT WHEN THE BOOK WAS PRINTED IN 1990.

TO ESTIMATE CURRENT PRICES, PLEASE KEEP IN MIND THE CONTINUING INFLATION, AND THE DRAMATIC INCREASE IN THE SELLING PRICES OF USED AIRCRAFT THAT IS TAKING PLACE.

TIPS ON BUYING CESSNA SINGLES

ISBN 0-916413-17-9

Printed in the United States of America

Aviation Book Co., 25133-E. Anza Dr., Santa Clarita CA 91355

TABLE OF CONTENTS

APPENDICES

CREDITS TO:

Steve Baisden of Pasha Publications for production work...Pasha Publications, publisher of defense industry newsletters, Arlington, Va., for use of production facilities...Joe Shagna of Maryland National Bank for the loan payment table...Don Kagle of the FAA Flight Standards District Office, Dulles International Airport, for help with all those airworthiness directives.

INTRODUCTION

Since this book was written in April 1988, interest rates have gone from 10% to 14.5% and down again, and the trend first reported here of aircraft sales to Europe has continued.

Also, it has become solidly a sellers market, meaning prices are skyrocketing. Prices increased dramatically until September 1988, paused (along with used aircraft sales) for two months, then continued upward.

The dilemma is this: if you do buy now prices are going to be well above *Blue Book* values. But if you don't, all the good used airplanes will be flying in Europe. How long do you have? Don't wait.

Looked at another way, if you buy as quickly as possible you may see your investment increase. But don't rush the sale. Take the time for a pre-purchase inspection.

This is an unusual book. It may help you to decide buying an airplane is the dumbest thing you, in your present situation, could do. If that is the result, the book is a success. Or it may indicate airplane ownership is a good idea. Either way, this book assures you make an informed decision.

I owned a Cessna 152 in leaseback for four years in the early 1980s, back when the more favorable tax situation was supposed to make leasebacks profitable. It didn't, but that is another story.

During the course of writing this book I came to several opinions about which airplanes are best to buy. Here we go:

(1) Get a 1983 Cessna 152, or newer, because that is the first year the engine was especially designed to cope with lead.

(2) Never buy a Cessna 172 with the "H" model engine. Skipping the "H" years means buying 1976, or older, or a 1981, or newer. Go for the 1981, because you get 160 hp.

(3) Never buy an airplane that has been run with auto gas.

(4) The 182 is probably the best buy in the Cessna fleet, but watch for in-flight carb ice and water in the bladder tanks -- unless the tanks have been corrected in the airplane you select.

(5) The Cardinal's resale value is a worry.

(6) One of the better engines in the general aviation fleet is the Lycoming 360 engine.

(7) The Hawk XP and Cutlass RG give you more maintenance costs from prop overhaul and retractable gear care, than benefits from speed advantage.

(8) If you are loaded with cash, both for purchase and operation, get the Cessna 210.

There is additional information in the appendices worth almost as much as the book itself. The FAA book "Plane Sense" explains what an airworthiness directive is, while the Cessna serial number list tells you such things as the horsepower of any model year you might purchase. The booklet alone costs $4.50 if purchased separately.

This edition of *Tips On Buying Cessna Singles* contains something unusual; a Piper service bulletin on inspections to prevent wing failure. Pipers are terrific airplanes, and the information is included to keep you from buying a bird with damage history, if you prefer Pipers to Cessnas. A used Warrior or Archer are great bargains.

Alton K. Marsh
Arlington, Va.

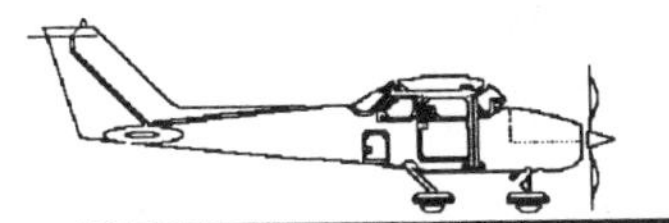
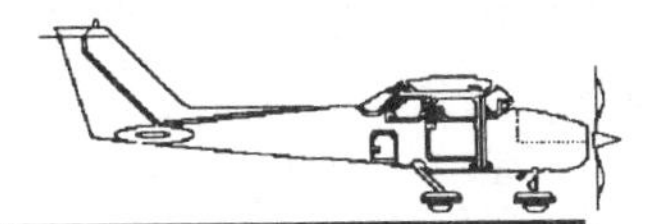

CHAPTER ONE

HERE'S WHAT IT WILL COST

How Much Can You Afford?

Most airplane buyers get their monthly payment in mind first, save 20% down, then go shopping for an airplane that fits the payments.

The chart below lets you pick the monthly payment you can afford, and then determine how much airplane that will buy. To do this, help was sought from Maryland National Bank. They have been in the airplane loan business many years. They also run the Air Power loan program of the Aircraft Owners and Pilots Association.

Obviously you can find dealer financing, as in

ESTIMATED AIRCRAFT MONTHLY LOAN PAYMENTS*

	Loan Interest Rate (Moves With Prime)				
Amount Borrowed	10%	12%	14%	16%	Term (Months)
$10,000.00	$212.50	$222.40	$232.70	$243.20	60
15,000.00	244.00	264.75	281.10	297.90	84
20,000.00	332.00	353.00	374.80	397.20	84
25,000.00	268.75	300.00	333.00	367.25	180
30,000.00	322.50	360.00	399.60	440.70	180
35,000.00	376.25	420.00	466.20	514.15	180
40,000.00	430.00	480.00	532.80	587.60	180
45,000.00	483.75	540.00	594.40	661.05	180
50,000.00	537.50	600.00	666.00	734.50	180
60,000.00	645.00	720.00	799.20	881.40	180
80,000.00	860.00	960.00	1,066.00	1,175.00	180
100,000.00	1,075.00	1,200.00	1,332.00	1,469.00	180
120,000.00	1,290.00	1,440.00	1,599.00	1,763.00	180
140,000.00	1,505.00	1,680.00	1,865.00	2,057.00	180
160,000.00	1,720.00	1,920.00	2,131.00	2,350.00	180
250,000.00	2,688.00	3,000.00	3,330.00	3,673.00	180

*Source: Maryland National Bank/AOPA . Requires 20% down. You must qualify for 180-mo. term.

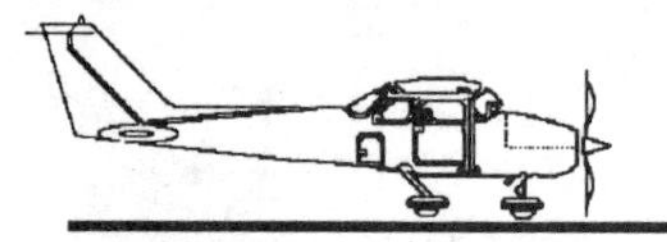
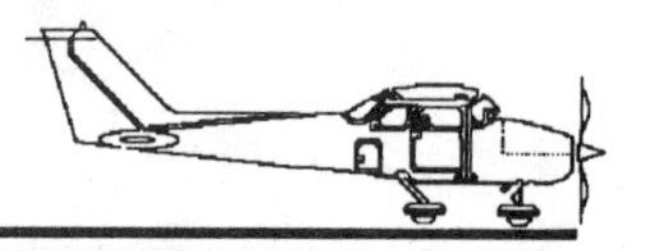

the case of Van Bortel and others. However, Maryland National Bank was asked first for information, and their loan officers cooperated fully.

The chart in this chapter tells you not only what your payment will be, but because the Air Power loan program has variable rates, it tells you what happens when the interest rates go up. Yes, you can also get a fixed rate, and a Cessna 172 club owner in Herndon, Va., recommends that approach to "...eliminate all the uncertainties you can."

The Air Power fixed rate is set at 3 to 3.5% above the prime rate at the time you take your loan. If you want to know what the prime is today, check the highest rate published in the Money Rates section of the *Wall Street Journal.*

The variable rate floats 1.5% above prime, adjusted at the end of each calendar quarter. Yes, you can start out with a floating interest rate now, to take advantage of lower rates, and convert to a fixed rate in the future if the economy gets sick.

The Air Power program also gives you — for a loan of $20,000 or more — free or reduced AOPA membership fees for as long as the loan is in effect. It costs $35 to join AOPA.

Finally, in addition to your dues, the Maryland National Bank/AOPA loan program will also perform the title search and FAA filing requirement normally charged to the buyer. Those fees are currently a minimum of $45.

When the loan goes into effect, the Maryland National Bank/AOPA loan program opens an extra, unsecured line of credit equal to 10% of the amount borrowed, up to a maximum of $10,000. That can be used for repairs, new radios, refurbishing or supplies. That might be necessary if you find a "surprise" waiting under the cowling of your newly purchased aircraft. The loan program includes amounts up to $500,000. Most customers, however, borrow about $30,000.

Join Cessna Pilots Association, AOPA Or The Cessna Owners Association

AOPA provides the greatest collection of services, in one spot, of all the places you can turn for assistance. For a measly $5, they'll send the AOPA buyers kit with lots of specific information and much general information on buying a plane. After that, for free, they will send you additional reprints of articles about other models you may be considering. The service is always fast and thorough, and you may even find yourself talking to a mechanic who has valuable insights. Naturally, this service is for *members*, so you need to join before expecting them to work for you.

Cessna Pilots Association

When you call the Cessna Pilots Association, as a member, they send you reprints of articles on the aircraft you are considering; but they don't have to. CPA official John Frank knows the information on Cessna models off the top of his head, and will give it to you over the phone. It is a great organization to have on your side. Like AOPA, the information is free to members. Their full color magazine is very well done, contains news from Cessna long before it hits the regular news media, and is filled with maintenance news you need as an owner, buyer or, for that matter, a renter. Phone 316-946-4777 to join.

Cessna Owners' Association

The Cessna Owners' Association is unique, in that it also operates a Piper Owners' Association.

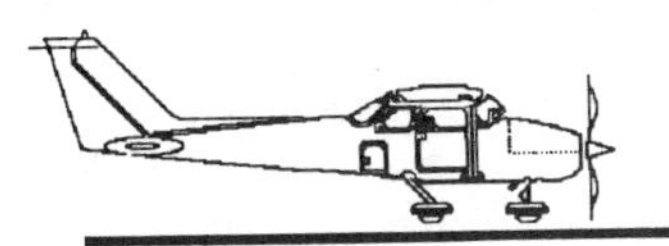
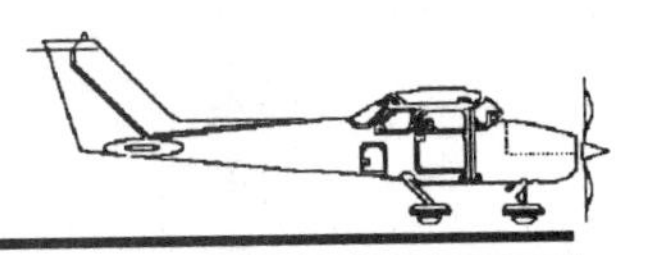

Call 1-800-247-8360 to join. Like the Cessna Pilots' Association, The Cessna Owners' Association also has full-color magazines for each of the organizations it operates. A phone call to someone like Judy Wood brings much assistance in purchasing a plane. First of all, she can look up any information you might need, like the day this author called wondering why he had never heard of a straight-leg Cutlass. Judy provided the answer: there are only 29 such airplanes registered in the U.S.

The organization even publishes a magazine of interest to corporate pilots, *Aviation USA*.

Another source of good information is the *Aviation Consumer*.

A recent issue pictured one of their staff shooting an "H" model engine on a Cessna 172 to put it out of its misery.

A Chat With Cessna Finance

There are of course other finance companies to consider besides the AOPA program. Cessna Finance was used by the author to buy a new 1982 Cessna 152.

But what a lot of rules Cessna Finance has! An official of Cessna Finance, who shall remain nameless and protected, said the problem comes from the fact that Cessna Finance is owned by General Dynamics, a rule-oriented aerospace company.

Cessna Finance cannot make loans of longer term than 96 months, but they do require only 10% down. It offers floating rate loans at 2% over prime, which in late December 1988 meant the loan rate was 12.5%. Their fixed rate on that date was 14%. In February 1989 the floating rate was 13.5% while the fixed rate had risen to 15.5%.

Cessna Finance will only offer more lenient rates or terms if it appears the local bank is going to beat them out of the business.

There are some other problems. Where AOPA will make a line of credit available to fix up the newly purchased airplane, Cessna Finance will only change the terms of the original loan. For example, Cessna Finance will only loan half the cost of an average engine overhaul. The overhaul you get might be higher than average, but Cessna won't offer more than half of the figure in its standard overhaul price book.

Cessna Finance also won't loan money for a new Piper, but that is understandable.

There is another point. Today numerous dealers are repainting, overhauling the engine and installing new interiors prior to the sale. But Cessna Finance will only loan the Blue Book value — the price of a plane that shows its age.

"We get numerous complaints about that policy, but we have to do what General Dynamics says," the official said.

And Cessna Finance will not take into account the higher prices of today's used airplane fleet.

Let's look at an example of a 1976 Cessna 172 that costs $26,000. Cessna Finance says the Blue Book value is $21,000, so that is all they will loan. The buyer must, (1) give the owner $5,000 out of pocket to make up for the difference between Blue Book and actual price, and (2) give the owner another $2,100 to satisfy the requirement of Cessna Finance to provide the owner 10% down. Now the buyer has spent $7,100. With another loan company, the buyer might have to put 20% down, but still needs a total of only $5,200 out of pocket.

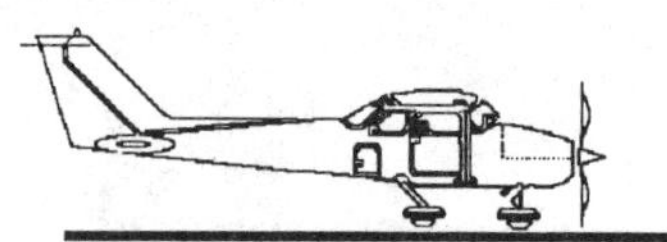
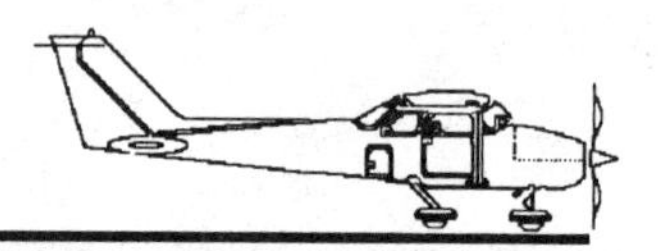

And another thing. Cessna Finance said in December 1988 that business was so good they were "restricting" the minimum loan to $25,000. They'll take the big profits, thank you. Admittedly, it's good business. But it's bad news for the purchaser of a $21,500 Cessna 172.

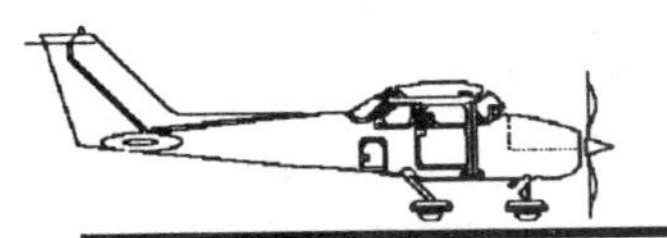
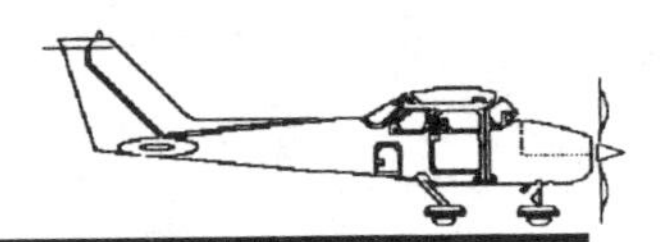

CHAPTER TWO

GO AIRPLANE SHOPPING!

The FAA's Own Advice

When you listen to Bill O'Brien you have to ask yourself, "Would I buy a used airplane from this man?" The answer is definitely yes, you would.

O'Brien is an FAA airworthiness inspector, by definition a nit picky sort of guy with a thirst for writing violations. But O'Brien goes one step further. He is an entertaining speaker who tells jokes on the FAA, himself and "bureaucrats."

For example, O'Brien defined FAA Headquarters, where he works, this way one recent evening: "The FAA Headquarters is like a log, floating down a white water rapids, bouncing off rocks, with 1,000 ants on board — each one convinced he is steering."

O'Brien used to be a mechanic, and knows those little secrets airplanes intentionally hide to create big repair bills. He has developed a method to inspect an airplane before purchase; he calls it the World's Best Preflight. It's not a bad preflight to use before every flight, considering the age of today's general aviation fleet.

The average single-engine aircraft is 21 years old. The average twin-engine piston was built 15 years ago. Turbine-powered twins average 11-12 years of age. Thus, the reason for a careful inspection.

First, The Paperwork

Boring, yes. Necessary, absolutely. Research, research, research your planned purchase, O'Brien says. Trade publications like *Trade-A-Plane* will tell you the asking price for the make and model you select. Also, get the operating handbook and start reading. Associations for specific airplanes, like the Cessna Pilots Association in Wichita, and clubs can provide valuable information. AOPA, the Aircraft Owners and Pilots Association in Frederick, Md., has such a list of organizations. If you are a member, call them and get the list over the phone.

Look at the FAA type certificate data sheet.

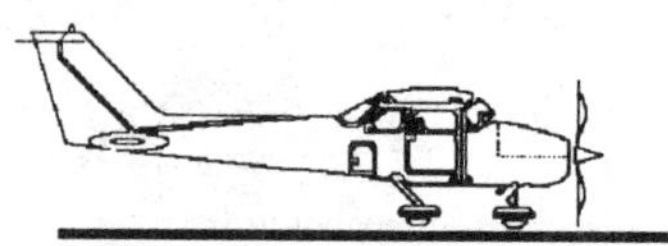

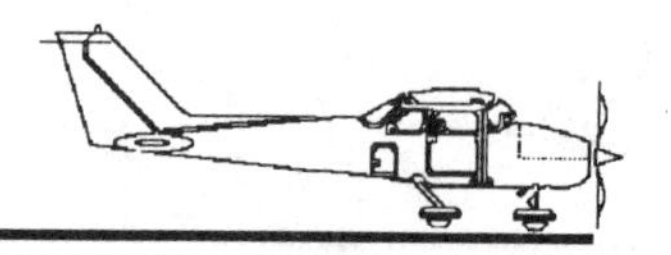

(We have included a type certificate summary for the Cessna 150/152, 172, 182 and 210 in this book.) It is the birth certificate of the airplane. Old airplanes like the Cessna 150 were built under the CAR 3 rule. The newer airplanes like the Beech Skipper, Piper Tomahawk, Seminole, Beech Dutchess and a few others are under FAR 23 certification. So what? Plenty, that's what.

First, the type certificate gives you a whole lot of information about the airplane — such as type of fuel, oil capacity, power limits, make and model of engine, etc. But what is REALLY important is the notes section. There may be mandatory retirement items listed there. What does "mandatory retirement" mean? Well, fuel and oil hoses must be "retired" in five calendar years on some aircraft built under FAR 23, like the Beech Skipper. Never heard that before? Even if the hoses only have 10 hours of flying time on them, they go off to the junk pile at five calendar years of age. The carry-through structure gets thrown away at 11,000 hours on one plane under FAR 23. When you throw away a part that important, you throw away the airplane. Cessna 150s never had mandatory retirement items.

Where do you get type certificates? Bigger fixed-base operators have data sheets, and the local FAA District Office has them. Each airframe, each engine and each propeller has its own data sheet. The FAA will have them on microfiche or paper copies.

The newer airplanes did not come out under FAR 23 until about 1978, although the regulation was in effect prior to that.

We're almost done with paperwork, but there is more to do when we get to the plane.

Flashlight, Mirror, Block Of Wood

Now it's off to the airport to meet the owner and examine his plane. Take along a flashlight, mirror, block of wood and current altimeter setting. O'Brien suggests a good tactic, if you would rather not do the inspection yourself, is to say, "I'll pay for a 100-hour inspection if you pay for the repairs." If the owner leaps in his car and fishtails down the road, don't buy the airplane. (Be aware you'll pay $500-$600.)

Make sure the airplane is in the same configuration as when it left the factory, or is in a configuration approved by the FAA. That configuration is listed in the documents already mentioned.

This will be the best preflight since your first solo. Since O'Brien was a mechanic at the time of his first solo, he nearly disassembled the airplane.

First, walk slooowly around the airplane. Slooowly, at a distance of 20 to 25 feet. You may find the owner has installed a rear view mirror to see aircraft during taxi. Messes up the aerodynamics and isn't approved. O'Brien found such an airplane at Manassas, Va., and called the owner with the dreaded words, "Hello, I'm from the FAA." O'Brien says it changed the owner's day. But all the owner had to do was take the mirror off the airplane. There was no further penalty.

Another airplane owner had electrical tape on the leading edge of the wing — apparently to avoid bug corrosion — but he taped over the stall strips on the leading edge in the process. The aircraft no longer had proper stall characteristics. O'Brien went to his phone again. "Hello, I'm from the FAA," he told the startled owner. O'Brien points out that if you buy the airplane, you are the one that can expect the call.

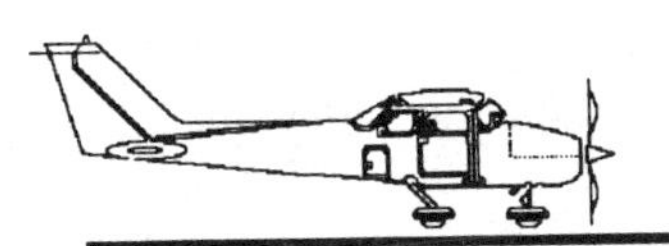
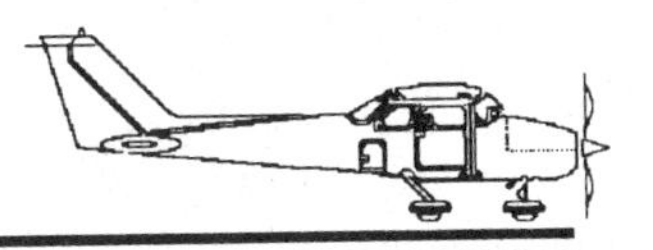

Now, take the airworthiness certificate out of the airplane's pouch and see if the "N" number on the fuselage agrees with the one on the certificate. Do the same for the serial number, the listed owner and other data. Now check the registration data. The airworthiness certificate is effective as long as approved maintenance and inspections are performed. That means you don't put rear view mirrors on the outside of the airplane.

After March 1985, all airplanes manufactured were required to have an approved airplane flight manual. The manual must have all the revisions. Some of them might include changes in stall speeds. The manufacturer or dealer will know what revisions are required.

Check weight and balance forms, and look at the date. Be suspicious if there are new radios in the airplane and no weight and balance information for them.

Don't fly any airplane unless you are sure the weight and balance is correct, O'Brien warns. An Aerobat owner wanted to put a larger engine on the aircraft, and moved the battery to the tail to make more room for the engine. Everything was fine till he literally went for an intentional spin, and found the battery weight in the back would not allow recovery from the spin. The aircraft crashed. The impossibility of recovery was later proven in tests using another Aerobat equipped with a spin chute.

Maintenance records must show, of course, the annual inspection and a statement that the plane is airworthy. Total time (not tach meter or other time) must be recorded. Current status of life-limited parts must be recorded. Time since overhaul of all items needing it must be required. Operating under FAR 91, the engine is not required to be overhauled! Bet you thought there was a regulation requiring engine overhaul. The engine manufacturer recommends it, and the FAA suggests the recommendation be followed, but it is not required. Failure to do so can cost big bucks later, however.

Look at the current status of compliance with airworthiness directives. A mechanic can — and probably should — help in reviewing that paperwork. If major alterations have taken place, there must be an FAA form 337 in the records. We're suggesting a pre-buy inspection here, of course. But a funny thing is happening to pre-buy inspections. Repair shops like one we called at Manassas, Va., in March, 1988, are afraid to do them. The reason is the same one that is crippling the entire general aviation industry: liability. The shops normally try to keep the cost down by limiting the hours of work to two or three, but in so short a time they could miss some major items. When the owner buys the aircraft, and discovers an expensive repair item, he heads for the shop that did the pre-buy inspection — leading his lawyer by the hand and shouting, "Get'em!"

The Airplane, At Last

Finally, we get to look at the airplane. It's about time. Oh, yes. One more thing. If the paperwork's no good, don't even look at the airplane. Go home.

There is an FAA bulletin that closely follows O'Brien's plane inspection tips. It is a yellow, stapled, three-page document numbered FAA-P-8740-15A. It is part of the accident prevention program, and is titled, "Maintenance Aspects of Owning Your Own Airplane." You'll find it in this publication.

Now then. After the paperwork passes muster, head for the aircraft's doors. Shut them and lock them. If they won't seat properly, rain could get into the aircraft. If they are distorted, you have

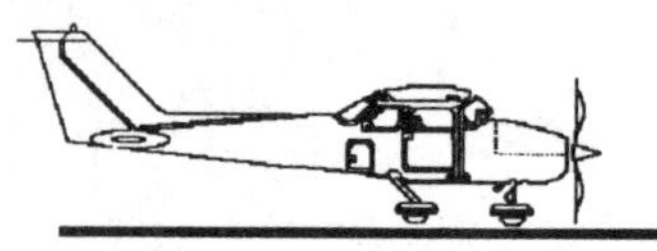

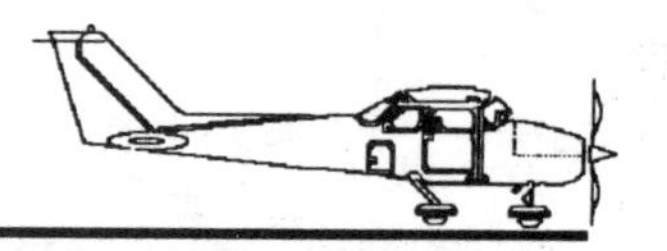

to ask, "What distorts a door?" If the door is distorted, the fuselage could be distorted. What distorts a fuselage? (Are you sure you want to know?) The answer, O'Brien says, could be a hard landing, high G loads in flight or someone leaving the doors open in high winds during preflight. Quite a tale that door tells, isn't it?

Look at the windows for crazing and delamination or a separation between layers. You can fly it, but it is unsightly and it will be difficult to spot other aircraft through it while flying.

Inspect the seat. When is the last time you inspected the seat prior to takeoff? Probably never, except for Cessna pilots who nearly push the rudder pedals through the firewall trying to find out if the seat rail will hold. Is the back going to flop down on takeoff? Is the seat going to slide backwards on takeoff? Holes in Cessna seat tracks are supposed to be round, not oval, by the way.

A student pilot in a Cessna 150 had a seat back failure and was thrown to the back of the airplane, restrained somewhat by the seatbelt. When his torso reached the back, the plane went out of trim and began a climb. The student ripped up the instrument panel trying to climb back uphill to reach the control column, but did so and made a safe landing.

"What the investigation report doesn't say is, he buried his underwear," O'Brien said. The incident might have ended there, except for one thing. Inspectors found the airplane lacked an annual inspection. It was time for another phone call to the owner. "Hello, I'm Bill O'Brien. I'm with the FAA," the call began. It went downhill from there for the owner.

As long as you are in the seat area, lift the rug and look for corrosion. If you find hardware (screws, washers), puddles of hydraulic fluid or rust, ask where they came from.

Make sure the fuel selector will turn off. Sometimes they are safety wired on. If you retard the throttle, you only shut the fuel off at the carburetor. But if a broken fuel line is squirting gas on the hot cylinders, you need to shut the fuel off at the fuel shutoff valve.

Check the instruments, as long as you are still inside the airplane. The airspeed indicator should say zero, for example. White haze inside the artificial horizon means vacuum system seals are leaking. Check the altimeter and record gas gauge readings. Then look in the tanks. The most important point here is that when the tanks are empty, the gauges should be accurate. It's an FAA regulation.

Directional gyros cost between $250 and $350 to repair. Turn coordinators cost about $250 to fix, O'Brien said. Check them in flight and during taxi, after they have had a few minutes to get up to speed.

Operating ranges should be painted on the tachometer and the tach should be accurate. Tachs that came as original equipment 21 years ago aren't accurate, O'Brien said. A study by FAA Headquarters proves it.

Check to see if the fuel primer has air in it. It shouldn't come out or go in too easily. If you have an airplane that has a good mag check at high RPM and a lousy idle, with black soot on the exhaust, the problem could be the primer, assuming there are no spark plug or compression problems, O'Brien said.

Move the controls smoothly to detect binding or restrictions of any kind. Binding will wear thin spots on control cables.

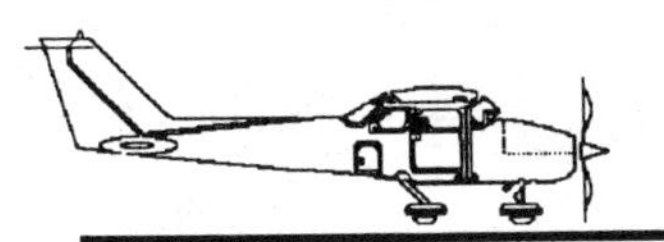
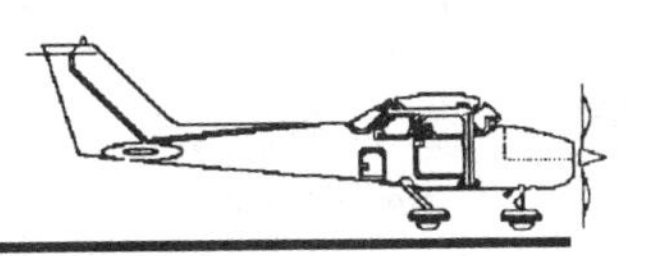

Look at proper installation of fire extinguisher mounts. If there are none, the hurtling extinguisher can kill passengers in gear-up landings that are otherwise so smooth that there is little other structural damage.

Check the rivet line on the wings. Look down the line like you were sighting down a rifle. If they are out of line, there could be a history of hard landings or a lot of accelerated stalls. Call a mechanic to check it.

Check screws on low-wing airplanes that hold the gear. They must be aligned, with screw heads pointing the same way. If they aren't, they are moving in flight. You won't be able to detect it on the ground because the weight of the wing is keeping them tight. You have steel screws working against soft aluminum, sort of like a diamond bit drilling through sandstone.

Check the rear spar area for damage from hard landings or tail strikes such as happens on trainers during soft field takeoff practice. Look for popped rivets or buckled aircraft skin in the tail area.

Fuel leaks are detected by stains and won't stop without expensive maintenance. "I'm going to stand here and tell you as an FAA — no, you wouldn't believe me. As an Irishman, I am going to tell you I have never met an airplane that healed itself," O'Brien said of fuel leaks. When you see fuel leaks and the owner said, "Aw, it's been like that for years," it means something happened somewhere down the line to cause the problem. And it will never "heal itself" unless you pay the healer (mechanic).

Take fuel samples — a lot of them — and take them after the test flight as well, looking for sediment, paint or water. The owner may be horrified by about the fifth cup of expensive fuel you throw away, but it beats throwing away money later.

Fuel caps which fit flush to the wing are more prone to leakage than those that stand above the wing. The cap sits down in the wing, and is constantly irritated every time the lineman refuels. O-rings on the cap wear, and can let water in. Pour one of those fuel samples on the flush cap. If it runs into the tank, the cap leaks. Fuel caps should not go "whoosh!" when opened. If they do, there is vacuum pressure building up in the tank that can prevent fuel from reaching the engine. The fuel vents aren't venting.

Rubber bladder fuel tanks like those on Cessna 182s and 210s can develop wrinkles that can hide over a gallon of water in little "lakes" between the wrinkles. The lakes of water are blocked from reaching the drain point where you take samples. Rock the wings to move the water around and take more samples.

Bird droppings can cause corrosion, and birds nests can leave material INSIDE the wing that can bind cables. In low wing airplanes, birds enter the wing — their high tech condo — through openings around retractable landing gear. Sometimes owners will tie up the controls with the seat belt, holding one aileron up, as the owners manuals sometimes suggest. Birds then enter from openings behind the raised aileron, tiny though the openings may be.

Here is another neat test. To check aileron alignment, put the flaps up and set aileron trim to zero. Then go out to the wing and pinch the aileron and flap together. They should be in trail. Now, look in the airplane and see if the control wheel is horizontal. Next, without touching anything, look at the other wing to see if the flap and aileron

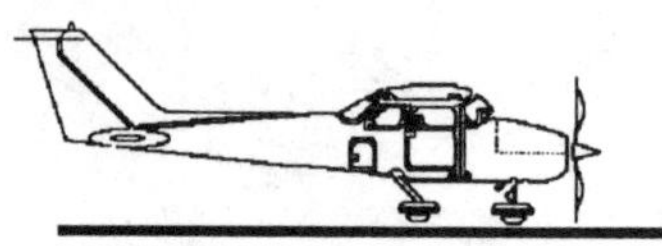
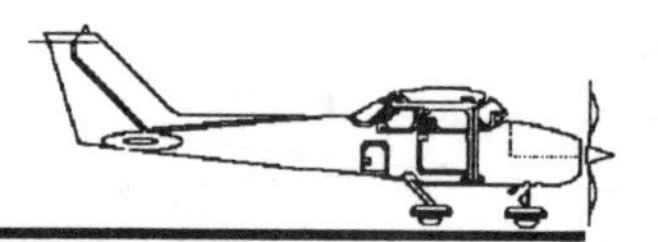

match up. If they are not within a quarter of an inch, you have a rigging problem, so sayeth the FAA.

Gear Problem Detection

If there is rust on the chrome portion of the landing gear strut, rust is coming through from beneath the chrome and a new chrome plating may be necessary, because often the struts will be out of production.

Dust covers should be on the end of the axles. If they are missing, dirt could enter the bearings and result in the wheel coming off.

"Just imagine you have been at the beach in a particularly rough surf and a lot of sand gets in your bathing suit. And you have to walk a half-mile to the shower," O'Brien told a safety meeting. The audience got the point.

Tires. Ever seen cracks in the sidewall? FAA regs say you cannot fly with cord showing or cracks in the sidewalls.

If tread on the inside of the tire is worn, the wheels are out of alignment. Don't just flip them over, as some owners do, to let the tire's other side wear. The wheels must be aligned. If the tread is worn smooth, there is one-sixteenth inch of rubber between you and an accident, O'Brien said.

To check brake linings, place a paper match against the lining. If there is still lining showing, you are ok. If not, replace the lining.

Look for metal fatigue cracks around brake bolts. Change brake linings every 100 hours and you won't have a problem.

Some owners will tell you Cessna nose wheels are supposed to shimmy. It's "... just a problem with Cessnas." Don't believe them. It could cost up to $200 to repair. Vibration could be caused by a broken dampener, loose dampener bolts or a nose wheel that is out of alignment.

The Engine

There are a few things you can check without being a mechanic. Look at the gap between the cowling and prop at the top of the spinner and compare it with the gap at the bottom. They should be equal. If not, the engine is sagging on its mounts. The rubber doughnuts that shock mount the engine to the airframe are old and have become compressed. They are expensive, and must be replaced, not switched or rotated with other rubber mounts. Obviously, while in the area, you will check the prop, spinner plate and spinner for cracks.

Look at paint stripes on the outside of the cowling to see if they line up.

Tap the air cleaner on white paper. If there is a lot of crud on the paper, there is a lot of crud in the engine. Look at the oil, and take a sample for an oil analysis. Metal in the oil screen means the airplane is telling you, "Don't buy me." Go home.

Burn marks or heat marks on cylinders could indicate pressurized gases escaping around the sparkplug hole. Sometimes cylinders will only show these markings when hot. Pretty blue haze on the exhaust pipe may mean the engine is running too hot.

Winter plates are not to be installed in warm weather because they make the engine run with super hot oil. If you fly to Florida in winter, take them off when reaching warmer temperatures. Slow damage to exhaust valves is the result.

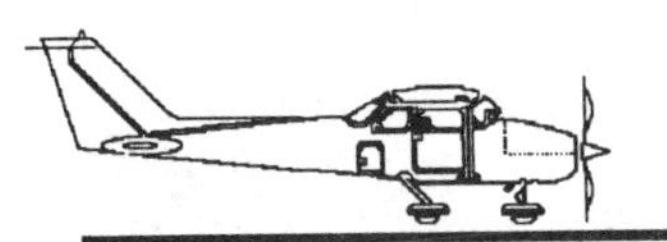
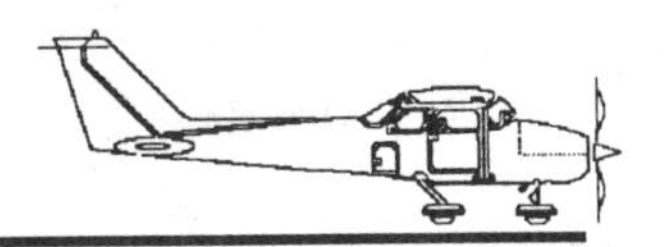

"Metal remembers," O'Brien said, meaning it remembers the damage you did to it once and will show it later. Blistered paint or a burnt smell around cylinders may also be an indicator of high temperatures.

Exhaust pipes should not touch at their open ends.

Duct or scat tubing — the kind with a wire coil in it that carries air around the engine — should not have holes. It costs $7 to $11 a foot.

There should be no openings in the firewall where exhaust gases could enter the cockpit. It goes without saying there should be no bends or buckles in the firewall.

If there is discolored exhaust, or the engine is hard to start, don't buy the plane.

Take the block of wood you brought with you and place it on the ground behind the prop. Now move the prop (Wait! Are the master and ignition off?) till it's back touches the block of wood. Now, move the other tip of the prop to the block of wood. It should be within one-sixteenth of an inch of the block.

Finally, look at the engine data plate and make sure it matches the one in the log book.

The Test Flight

You might want to have an instructor come along if you do not know the owner. Do precision maneuvers. Run everything to see if it works. Everything in the front office should work. (One dealer we know said you can always tell if a buyer is "green" because "They want to fly it." Our suggestion: be green. Insist on flying it to discover rigging and other problems.)

Before takeoff, put the controls in neutral and see if that is where the control surfaces really are. Check the ELT five minutes past the hour for three sweeps.

Check control rigging in flight by taking your hands off the airplane, once it is in trim, and seeing what it wants to do. By the way, a good preflight test of rigging is to wiggle the rudder side to side with your fingers. If rudder cables slap against their guides, the cables are loose. That is difficult to do with some types of Piper nosewheel steering, but not difficult with a Cessna 172.

After the flight test, have an expert look at the paperwork — a mechanic who is familiar with the type of aircraft you want to buy. As we said, you may have to hunt for a mechanic that is still willing to assume the liability for a pre-buy inspection.

The mechanic will be more willing to do a 50-hour inspection and a compression test. It is a good idea, besides.

The final things you need are a title search and a quick calculation of the life left in the engine. Subtract the time on the engine from the recommended time between overhaul. That is how much flying you are buying. If there are 1,500 hours on the engine and the time between overhaul is 2,000 hours, then you are buying 500 hours of flying. After that, it is off to the overhaul shop, where prices start at $5,000.

Next comes a bill of sale and a registration form, needed to keep the FAA happy, and presto; you're an airplane owner!

One final tip. To help with the avionics check, the FAA has bulletin number FAA-P-8740-18, "Preflighting Your Avionics." It is an excellent guide to a tough preflight for avionics that matches well with O'Brien's tips for the overall aircraft. While he didn't mention it, it was handed out during one of his talks. We put it in the back of this manual, also.

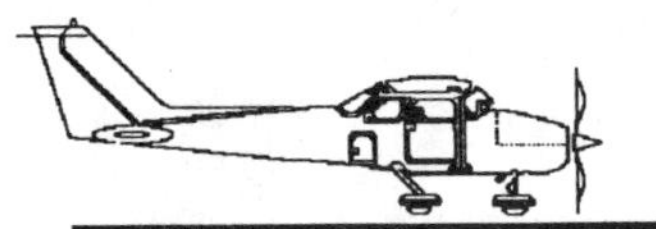
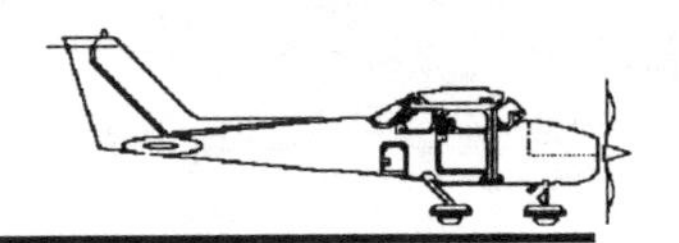

WORLD'S BEST PREFLIGHT

(You will need: flashlight, mirror, block of wood, current altimeter setting.)

GENERAL INFORMATION

Owner ___________________________ "N" Number ______________________________

Address ____

Phone ___________________ Asking Price____

A/C Make and Model_____________ Serial Number________________________________

Engine Make and Model ____

Engine Serial Number____

Overall Condition of the Paint (Grade 1 to 10) _____________________________________

Overall Condition of the Interior (Grade 1 to 10)____________________________________

In general, do you want newer avionics than what it has?____________________________

OVERALL INSPECTION

(1) Walk slowly around the airplane at a distance of 20 to 25 feet. Are there any unusual modifications, such as electrical tape on the wing leading edge, special wingtips or rearview mirrors? What about nicks, dents and scrapes in the wing, wingtips and fuselage? Is the skin buckled, especially at wing roots?

OK ___________ Problems ____

____

____

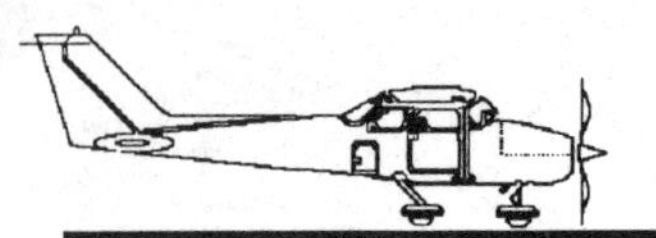
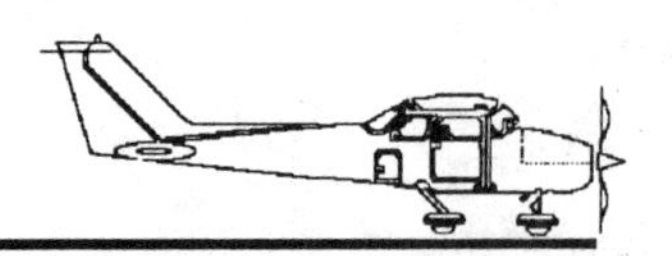

THE PAPERWORK

Remove the airworthiness certificate, registration, radio station license, weight and balance form, equipment list and maintenance records.

(2) Do they all show the same make, model, serial number and registration number? If not, go home.

(3) Is the "N" number and the serial number on the airplane data plate the same as the one on the above certificates?

(4) Are the engine and propeller model and serial numbers on the data plates the same as those in the maintenance records?

(5) Date of the weight and balance form. ________________________ Are there new radios in the plane, or any new equipment that is not listed on the weight and balance form? It may not be flown again until the weight and balance form is correct.

(6) Look at the maintenance records. Copy down the following information.

Date of Last Annual Inspection. ___________________________

Total Time on the Airframe. _________________ hours

Time on the Engine. _______________________ hours

When Does Manufacturer Recommend Engine Overhaul? _______

Time on the Propeller. _____________________ hours

Major Repairs and Alterations:

____

____

____

____

____

____

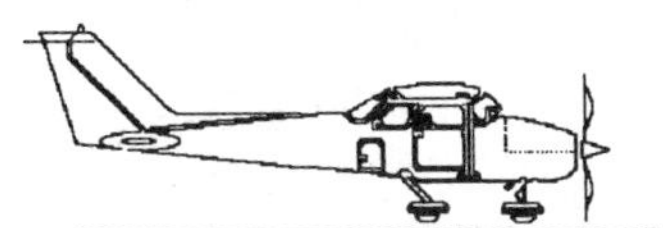

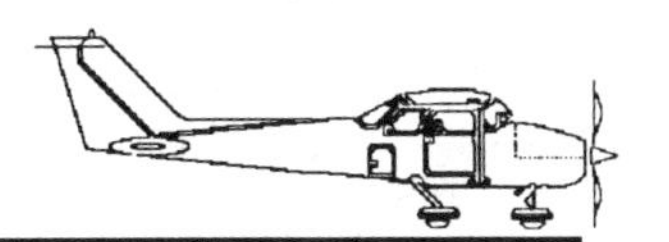

DETAILED INSPECTION

(Do these steps in addition to a routine preflight.)

Cockpit

(7) Open and close all the doors. Do they seat properly? Is the rubber gasket around the door undamaged?

OK ________ Problem ____

(8) Check all windows for crazing and delamination. Are they clear or "foggy"?

OK ____________ Problem ____

(9) Get in and sit in the seat. Does it need new covering? Is the back going to flop down on takeoff? See if the back locks and the seat locks on the seat rails by pushing on the rudder pedals.

OK __________ Problem ____

(10) As long as your feet are on the rudder pedals, are the brakes spongy?

OK ________ Problem ____

(11) Get out and look under the rug. Is there corrosion or loose hardware? If there is hardware, where did it come from? Is there rust or hydraulic fluid?

OK ______ Problem ____

(12) Move the fuel selector through each position and check for ease of operation. (Some are wired to the "on" position.)

OK______ Problem ____

(13) Check the instruments. The attitude indicator and heading indicator faces should be checked for white haze or fog on the glass. This is an indication of killer moisture inside.

OK __________ Problem ____

(14) The airspeed indicator should say zero and the altimeter should indicate approximate field elevation when set to the current altimeter setting.

OK ____________ Problem ____

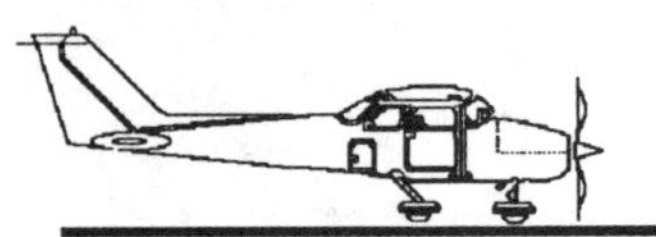

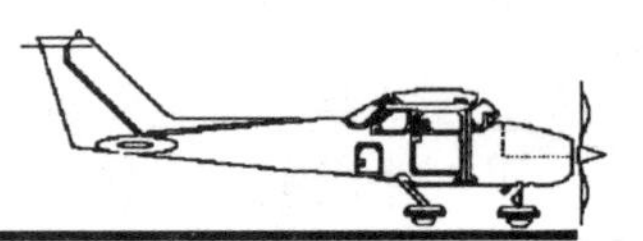

(15) Check the gas gauges. Now, look in the tank and see if that is really what the indication should be.

OK ________ Problem ______________________________________

(16) Operating ranges should be painted on the tachometer.

OK _______ Problem ______________________________________

(17) Pull the primer out, let it fill, and push it in. It should not go in or out too easily. If it does, it could have air in it.

OK ______________ Problem ______________________________________

(18) Move the controls smoothly to see if there is any binding or restrictions. Listen for sounds of binding.

OK ___________ Problem ______________________________________

(19) Check fire extinguisher mounts. Make sure they will hold in rough air or a rough landing. If there is no fire extinguisher, why not?

OK _________ Problem ______________________________________

Fuselage

(20) Look down the rivet line on the wings, like you were sighting down a rifle. Are they out of line? (On low wing airplanes there is a set of screws above the landing gear, visible to the pilot, which hold the gear on. The heads of those screws should all be turned the same way.)

OK ___________ Problem ______________________________________

(21) Check aileron and controls alignment. Put the flaps up and set aileron trim, if there is any, to zero. Go out to the wing and pinch the aileron and flap together. Now look in the airplane and see if the controls are neutral. Next, go to the other wing to see if the aileron and flap are within a quarter of an inch of one another (while remaining exactly aligned on the opposite wing.)

OK _________ Problem ______________________________________

(22) Check the rear spar area at the tailcone for damage from hard landings or tail strikes. Any popped rivets or buckled skin?

OK ____________ Problem ____

(23) Shake the rudder to determine if the control cables make slapping, whipping sounds. If so, they could be too loose and out of rig. (Careful not to hit the control stops.)

OK ____________ Problem ____

(24) Check antenna attach points for security. Are any broken?

OK ____________ Problem ____

(25) Look for fuel leak stains underneath the wings. Take four fuel samples from each tank to look for rust, other sediment, water or paint.

OK ____________ Problem ____

(26) Fuel caps should not go "whoosh" when opened, indicating they have built up a vacuum. Use a flashlight to look into the tank for sediment. One owner found a fuel intake port on the bottom of the tank. Fuel caps should not be worn.

OK ______________ Problem ______________________________________

Nose Area

(27) Now for that block of wood you brought. With the master and ignition off, put the prop in the 12 o'clock to 6 o'clock position. Place the block of wood on the ground just behind the tip of the prop. Now move the prop through a half circle, stopping again in a 12 o'clock to 6 o'clock position. Does the tip still touch the block of wood? It can be only one-sixteenth of an inch off.

OK __________ Problem ____

(28) Look at the gap between the prop spinner and the cowling. Is it equal on top and at the bottom of the spinner? If not, the engine mounts are sagging.

OK ____________ Problem ____

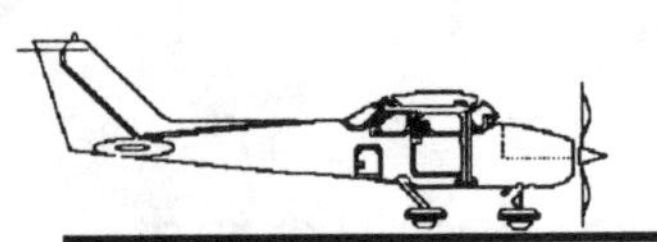
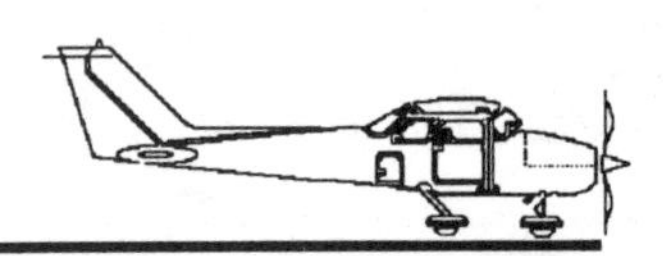

(29) Is the spinner cracked? Is the prop cracked or cracked and painted over? Are there nicks that need to be removed? What about corrosion and alignment?

Spinner: OK __________ Problem ______________________________________

Prop: OK __________ Problem ______________________________________

(30) Look at paint stripes on the outside of the cowling to see if they line up.

OK ____________ Problem ______________________________________

(Now, remove the cowling.)

(31) Tap the air cleaner on white paper. Dirt in the filter means dirt in the engine.

OK _______________ Problem ______________________________________

(32) Look for burn marks on cylinders. If the engine has recently been spray painted, maybe it was to cover up a problem. Is there blistered paint or a burned smell?

OK ____________ Problem ______________________________________

(33) Do the exhaust pipes touch? Is there a blue haze on them indicating the engine has been operated too hot?

OK ______________ Problem ______________________________________

(34) Is the firewall buckled? Are there openings in it where exhaust gases could enter the cockpit? Also, the winterization plate should be removed in warm weather

Firewall: OK _____________ Problem______________________________________

Winterization Plate: OK ____________ Problem ____________________________

(35) Look at scat or duct tubing around the engine. It should have no holes.

OK _____________ Problem ______________________________________

(36) Now, replace the cowling and start it up. If it is hard to start, or there is discolored exhaust, go home.

OK ____________ Problem ______________________________________

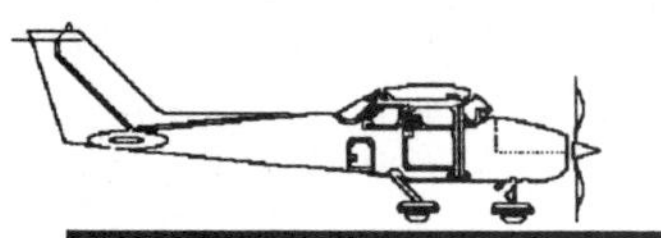

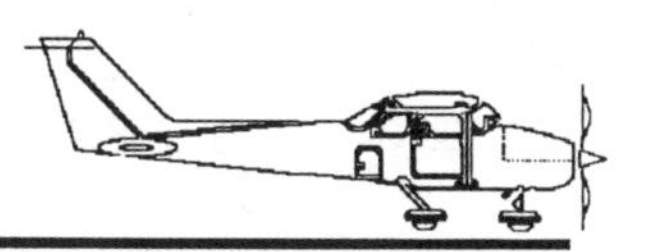

Landing Gear

(37) Is there rust coming through the chrome portion of the nose gear strut? If so, parts may be hard to find and chrome replating is expensive.

OK ____________ Problem ______________________________________

(38) There should be a dust cover on the ends of main gear axles.

OK ____________ Problem ______________________________________

(39) Tires cannot have cracks in the sidewall. Such aircraft are prohibited from flying.

OK ____________ Problem ______________________________________

(40) Tires should not wear more on one side than the other. If so, the wheels are out of alignment.

OK ____________ Problem ______________________________________

(41) To check brake linings, place a paper match on the brake pad. If there is still pad showing, it is adequate.

OK ____________ Problem ______________________________________

(42) Look for fatigue cracks around brake bolts.

OK ____________ Problem ______________________________________

(43) Cessna nose wheels are NOT supposed to shimmy, no matter what the seller may think. Does it? You may have to wait for the flight test to check this item.

OK ____________ Problem ______________________________________

FLIGHT TEST

(44) Before starting the engine, put the trim in neutral and see where it actually is.

OK ____________ Problem ______________________________________

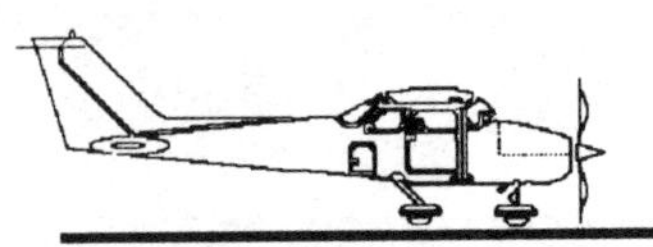
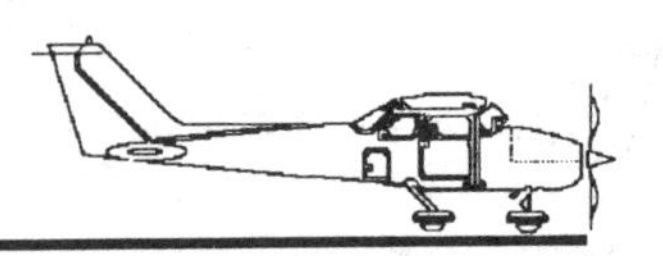

(45) Try all the radios and the autopilot to make sure each works and is within tolerances.

Nav/Com #1 ____

Nav/Com #2 ____

ADF ____

Glideslope ____

Transponder ____

Other Radios (DME, Loran) ____

Autopilot ____

(46) What about airspeed? Is it as fast at normal cruise as the manufacturer promised?

Cruise Airspeed ____

(47) While climbing at the recommended airspeed, is the vertical speed indicating what the book promised?

Feet Per Minute Climb ____

(48) Does the heading indicator have excessive precession? Is the turn and bank indicator showing a two-minute turn in two minutes?

Heading Indicator ____

Turn and Bank ____

(49) Take your hands off the controls. Does it stay level? Now take your feet off the rudders. Does it yaw?

Rigging of Controls ____

Rigging of Rudder ____

(50) Check the ELT five minutes past the hour for three sweeps.

OK ______________ Problems ____

Now, call the mechanic for a pre-buy inspection, and see if he agrees with you.

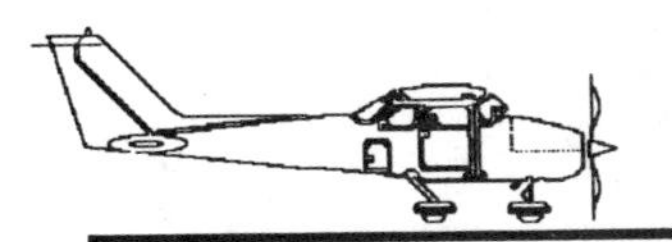
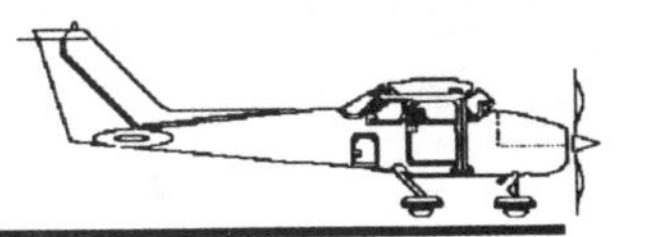

Oil Analysis Companies Recommended By The FAA

Aviation Laboratories

Write their main office at 3120 White Oak Dr., Houston, TX 77007, for their basic $12.50 analysis kit. Their phone number there is (713) 864-6677. They also have laboratories at Los Angeles International Airport, 12911 Budlong, Gardena, CA 90247 (phone (213) 217-9369), and at New Orleans International Airport, 918 Maria St., Kenner, LA 70062 (phone (504) 469-6751). They do all the additional tests as in the basic kit below, but each is priced individually.

Analysis, Inc. Of New Jersey
61 Suttons Lane Piscataway, N.J. 08854
Phone: 201/985-8282

Analysis believes you should have all the information possible, and includes a lot more tests that the kit above, which only tests for wear metals. This test costs $25.35. You must contact the company for a sample kit which consists of a four-once sample bottle and instructions. This test covers viscosity, acid number, fuel and water pollution, solids (carbon) and 21 elements, such as iron, lead, copper, chrome, aluminum, etc.

The Title Search

The easiest thing to do is phone AOPA's title search company (1-800-654-4700). That's what was done just to give you an idea about title search costs. However, if you want to try to find something cheaper by shopping around, by all means do so.

The AOPA Aircraft and Airman Records Department in Oklahoma City, Okla., has a standard package for $115 that includes a rush title search, a search of all the Airworthiness Directives and a search of any accidents involving your selected airplane.

Just an ordinary title search, however, costs $35, with a $5 charge for rush orders.

Here are the types of services most title search companies will offer. This particular listing is from the AOPA firm. All charges are higher for non-AOPA members.

— Title Search. A title search of FAA aircraft records to determine ownership of the aircraft and a list of any existing liens recorded by the FAA. $35 for AOPA members. $45 for others.

— Airworthiness Directive Service. A computer search is made based on your aircraft's serial number. You will get a printout of all airframe and engine ADs. Also, service letters and bulletins are listed. You take the list of identification numbers and compare them to the engine and airframe logbooks to confirm all ADs have been complied with. $50.

— Service Difficulty Reports. A printout of all reported service problems for your make and model aircraft for six previous years. The problems may have been submitted by pilots, mechanics, inspectors and maintenance shops. Prices quoted individually.

— Aircraft Incident/Accident Reports. You can either request this report for a specific airplane, or a generic make and model. Prices quoted individually.

— Import/Export Clearance.

— Airmen and Medical Data Service.

— Customized Services. AOPA can arrange

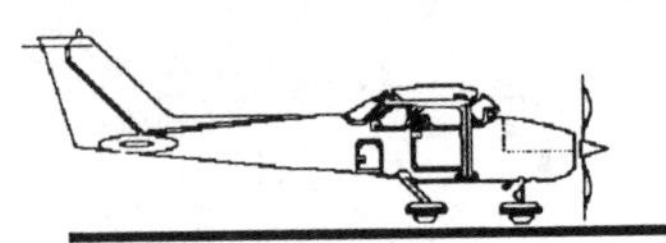

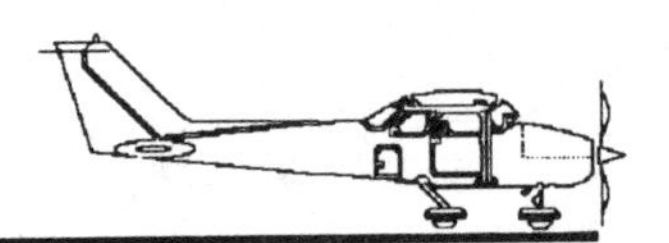

for all parties to be brought together physically or by conference call to close on the aircraft sale. When complete, the vital documents are filed for you, and your recording report is time-stamped at the FAA and forwarded to you.

What Cessna Says

Cessna supplies a supplement to its owners manuals called the "Pilot Safety and Warning Supplements." It contains a lot of useful information probably aimed at reducing liability for Cessna in court as well as helping the pilot/owner. In it is a section called "Airworthiness of Older Airplanes." It is useful to note some of the tips contained on those two pages.

First, Cessna notes that service experience reveals the aging airplane needs more care and special attention, and sometimes more frequent inspections of structural components due to wear and tear and exposure to the environment. Typical areas needing more frequent inspection include:

— Wing attach points and fuselage carry-through structures;

— Wing spar capstrips, especially the lower ones;

— Horizontal and vertical stabilizer attach points and spar structure;

— Control surface structure and attach points;

— Engine mounts and cowlings;

— Landing gear structure and attach points;

— Structural and flooring integrity of seat and equipment attachments;

— Pressurized structures (not a problem on the Cessna 172), especially around all doors, windows, windshields and other cutouts on pressurized airplanes.

Engine and related components that Cessna says should get special attention — with special attention to overhaul and replacement intervals — include such items as filters, hoses, actuators and engine accessories.

Cessna also warns that corrosion, overloading or damage to structure can drastically shorten fatigue life. Piper is facing some real problems now with operators who give rough treatment to single-engine planes, landing them off airport in Alaska or flying them at engine redline speeds in turbulence at 200 feet while on pipeline patrol. Cessna, while not facing the wing spar problems such treatment has caused Piper aircraft (see appendix G), has this to say:

"Be alert to the possibility that the airplane is being used in a manner different from the originally intended mission profile. Low altitude operation such as pipeline patrol, sightseeing, or **training operations**, will subject airplanes to more fatigue damage than high altitude cruise."

Other tips offered in the supplement fall into the category of, "Please do what Cessna says." But the firm does offer this final, interesting piece of advice:

"The prices of older airplanes tend to be relatively low when compared to newer ones. A significant portion of the reason for this is the much higher maintenance costs of the older airplanes. Be careful, when purchasing a previously owned airplane; the 'bargain buy' may need an extraordinary amount of maintenance to keep up with ordinary wear and tear, plus mishandling and lack of maintenance by previous owners."

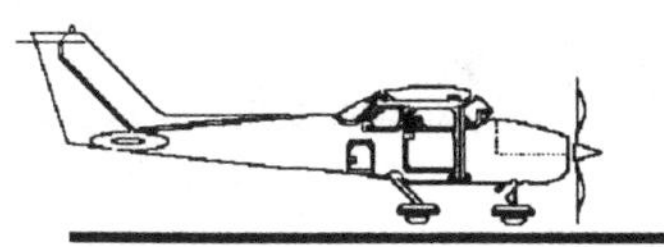
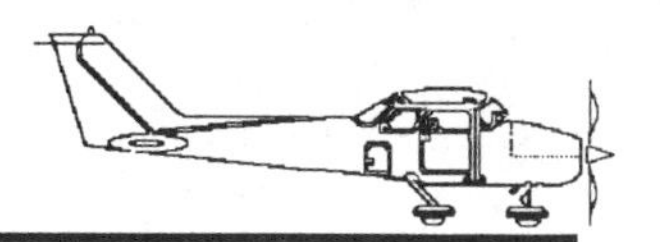

CHAPTER THREE

THE CESSNA 172

Now that you know how much you want to spend, how much airplane will that buy?

To find out, an analysis of two issues of *Trade-A-Plane* was done. The first issue of *Trade-A-Plane* showed 76 Cessna 172s available. It was one of the biggest listings among all the aircraft advertised.

However, the second March issue had 300 Cessna 172s advertised!

During analysis of the first *Trade-A-Plane* issue, a hypothetical buyer was created who did not want to make payments on any sum larger than $23,000, and would put 20% down. The analysis revealed the buyer could get a 1977 Piper Warrior with 2,000 hours total time and 1,300 hours since major overhaul. Or he could have purchased a 1977 Cessna 172 with nearly 2,200 hours on the airframe but only 1,000 hours since major overhaul. The Cessna would have given more hours of flying than the Piper before another overhaul (at 2,000 hours) was required.

Now, however, the model "H" engine factor comes into play. The 1977 Cessna 172 had the "H" model Lycoming 0-320 engine, making it the riskier buy. The best alternative for our buyer then became a 1976 Cessna 172, and for the amount the buyer had to spend, he could get a very nice 1976 model.

Let A Few Hours Pass After Overhaul

Let's take a moment for some tips before looking at those 300 172s offered in March.

(1) Look for an airplane that has 500 hours since major overhaul, unless, of course, the engine is factory new. The reason for that is, as a Manassas, Va., Cessna 150 owner once put it, "You don't want to be the first to fly a recently overhauled engine. You spend the first 500 hours wondering if there was anything wrong with the overhaul."

The owner said it was his opinion that the best place to have an overhaul done is Mattituck. If the engine is overhauled by Mattituck less than 500 hours before the sale, go ahead and buy it. Mattituck officials said they return the engine to factory specs.

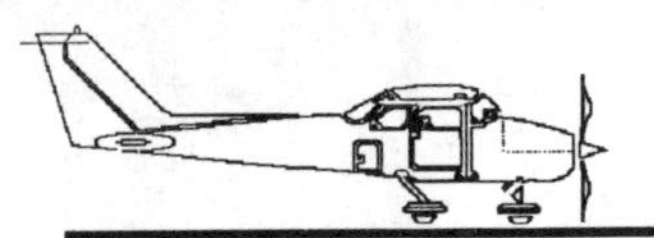
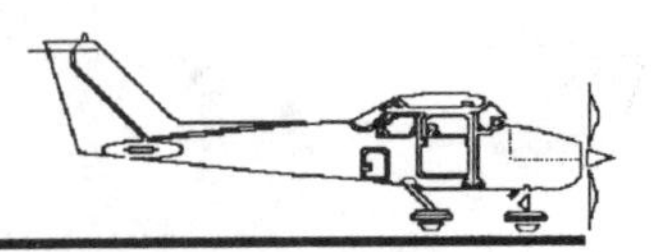

(2) You also don't want more than 2,500 hours on the airframe. It is the practice of a well-known, top quality Piper dealer to stock only used aircraft with 2,500 hours or less. In Cessna's case, even the Cessna owner's manual supplement warns that problems can develop as any airplane ages, although the Cessna wing is probably the strongest of any on the market due to use of wing struts.

So it makes sense to have a total time in mind as a goal, say a Cessna 172 with no more than 2,000 to 2,500 hours total time, or a comparable Piper with no more than 2,000 hours. The reason for the smaller number of hours for Piper is the concern seen in the past year over single-engine Piper wings that have seen rough flying. How do you know whether a plane has seen rough service? There's no logbook for numbers of river bed landings in Alaska, or numbers of hours spent at 200 feet, full throttle, on pipeline patrol.

(3) If you have the money, get an aircraft that is no more than nine years old. There are simply fewer maintenance problems on newer aircraft.

Here's What You Get

The analysis of those 300 Cessna 172s revealed these average costs:

— $8,000, a 1956 model, probably VFR;
— $12,000, a 1960 model, probably VFR;
— $13,000, a 1965 model, possibly IFR equipped;
— $19,500, a 1970 model definitely IFR equipped;
— $24,000, an excellent, 1976 IFR airplane;
— $21,000, a 1977 model, cheaper because it is the first of four years of the problem engine

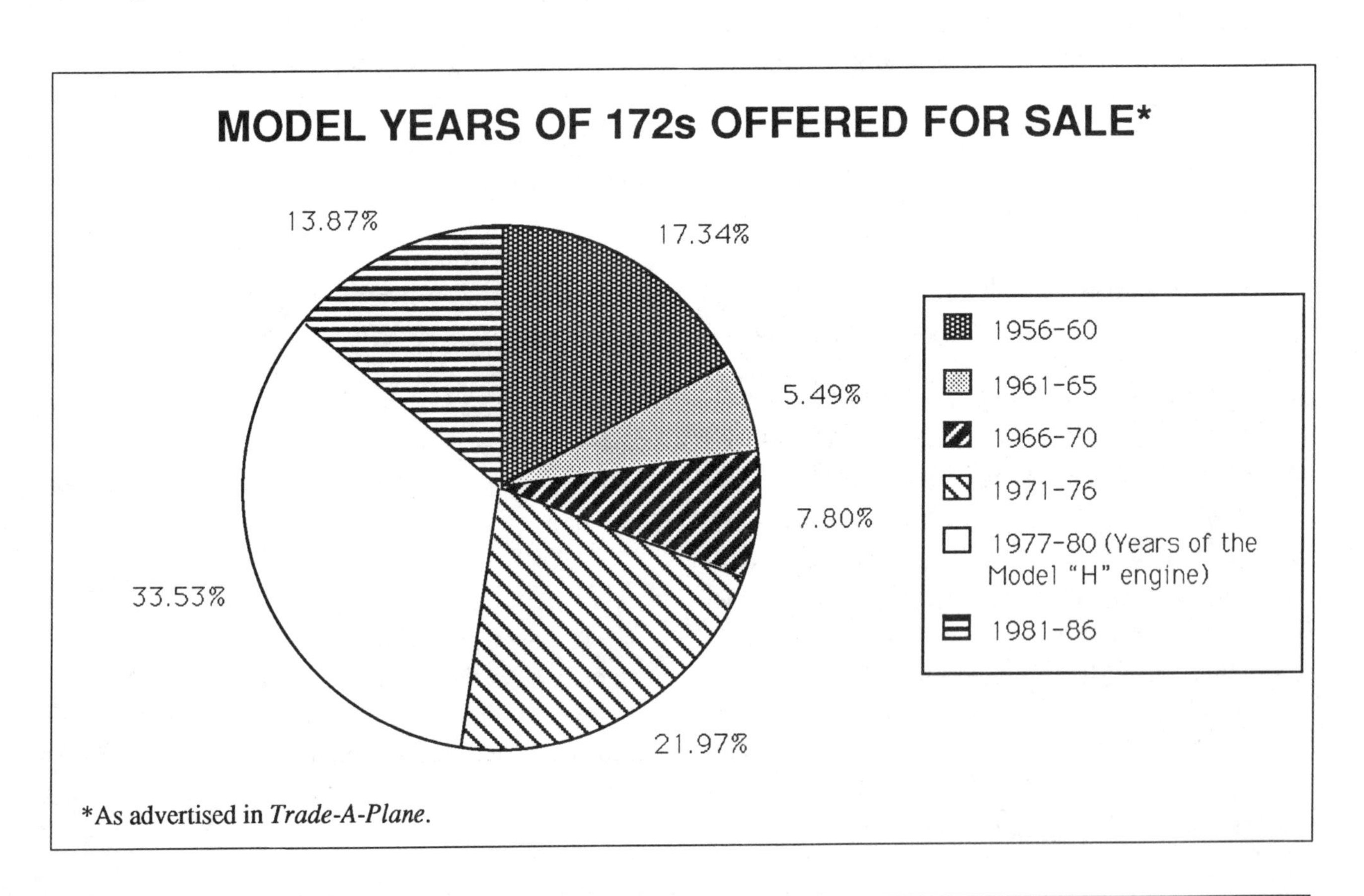

*As advertised in *Trade-A-Plane*.

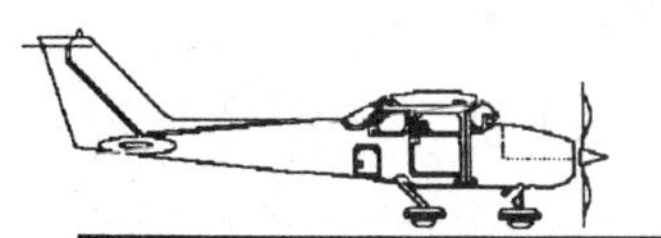

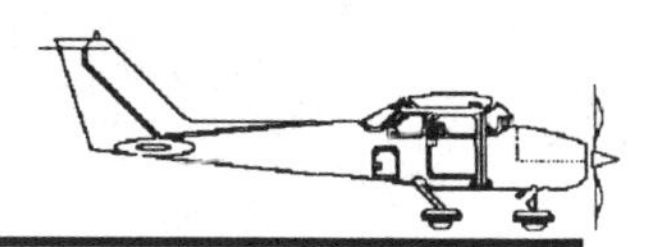

(Lycoming 0-320-H2AD), IFR;

— $29,000, an excellent 1980 IFR aircraft, but still with the "H" engine;

— $34,000, serious money but it gets you a clean 1981 IFR model;

— $43,000, an excellent quality 1983 IFR model, so close your eyes and spend;

— $54,000, top quality 1985 IFR machine (a nearly new one with less than 200 hours on it was advertised for $75,000);

— $60,000, top of the used Cessna 172 line, IFR of course.

Those are all "average" prices. You will find prices a few thousand dollars above and below each of those, depending on the circumstances of each plane. Those circumstances include: time on the engine, time on the airframe, quality and amount of avionics, status of the paint job and quality of the interior.

By far, the largest segment of 172s advertised were for the years of the problem engine: 1977-80. There were 116 airplanes in that category compared to: 60 for the period 1956-60, 19 for the years 1961-65, 27 for model years 1966-70 and 76 airplanes for the years 1971-76. There were 48 newer airplanes— those built between from 1981 to 1986.

The bottom line: even if you throw darts at an open copy of *Trade-A-Plane*, you are most likely to hit a model from production years 1977-80, and those are the ones with the Lycoming 0-320-H engines that gave Cessna so much bad publicity.

How To Read Those Ads

When you spend two days reading each of the ads for those 300 172s, you begin to pick up on little tips in the ads that serve as warnings, as well as tips revealing which are good. (In addition, your eyesight begins to blur, since *Trade-A-Plane*

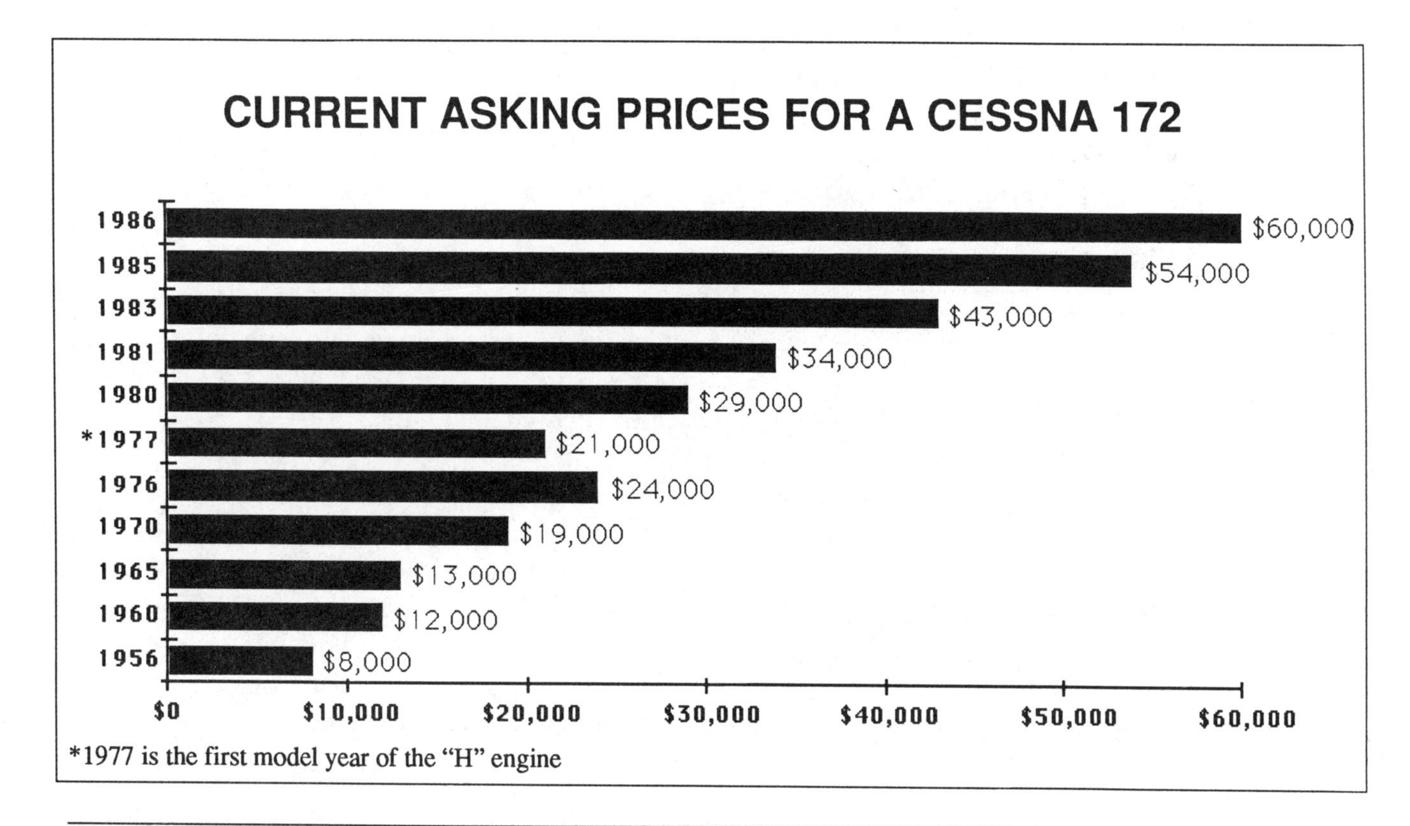

*1977 is the first model year of the "H" engine

type is small.)

Some of the ads for the model years which had the "H" engine were especially interesting. Many of the aircraft had high time, original engines, meaning they experienced no problems. Others revealed 1,900 as the total time of the airframe, but only 900 hours on the engine, meaning the engine needed a midtime overhaul at 1,000 hours. Some of the aircraft had engines from earlier model years, meaning the owner decided to dump the "H" model. Still others had conversions to 180 hp engines, partly because the owner wanted a more powerful engine or possibly because the owner thought it was a partial solution around problems with the "H"model.

Other ads for the 1977-80 models indicate their "H" engine had the "T-mod" done, meaning a mechanical modification was performed to help provide better lubrication of the engine. But the T-mod didn't really cure the problem. Using the oil additive properly was the best cure, but some owners didn't know what to do.

A few ads for those model years said nothing at all about the "H" engine, leaving the impression that if the buyer doesn't ask, nothing will be volunteered. Many, however, added notes such as "All ADs complied with," to let buyers know the engine problems were addressed.

You Can Find An IFR 172

The majority of the 300 aircraft advertised are equipped for IFR flight. If you want to save all the money possible, then get an older airplane that is VFR only. But it will be difficult to find a newer 172 on the used plane market that is VFR only. A surprising number are equipped with King avionics, usually preferred by Cessna owners to the avionics that come with the airplane. A few even have Collins, the top of the line in the opinion of many.

Obviously, high-time airframes are going to have a lot more trouble than those with fewer hours. Many had 5,000 hours on the airframe, and that is a lot of engine vibration and landings.

Blessed are the ultra-honest ads that give a fair assessment of the aircraft, pointing out the problems that a lot of flying can cause.

Watch those aircraft with long-range tanks. Extra fuel means fewer passengers and limited baggage. A Cessna 172 is normally a full-fuel, pilot-plus-two-passenger airplane in the first place. If you want to carry three passengers and baggage in a plane with long-range tanks, you will constantly be leaving the tanks half full, and inviting buildup of water in the tanks.

DME Is Worth The Search

DME is terrific to have on an airplane, and many of the used aircraft advertised have it. See if you can hold out for one of those. By the same token, encoding altimeters are almost a necessity, whether you believe they should be required by the FAA or not. It is the way of the future, and lots of used 172s have them.

A surprising number of the 300 aircraft had loran. But just any loran isn't a good buy. Some of the older ones require a degree in computer science to operate. (Ever read the Texas Instruments loran instruction manual?)

A half dozen 172s advertised had air conditioning, but that robs you of passenger and baggage carrying ability, which isn't all that great even without the extra equipment. However, if you operate out of Death Valley International, be aware that you can find a used Cessna 172 with air conditioning if you really look.

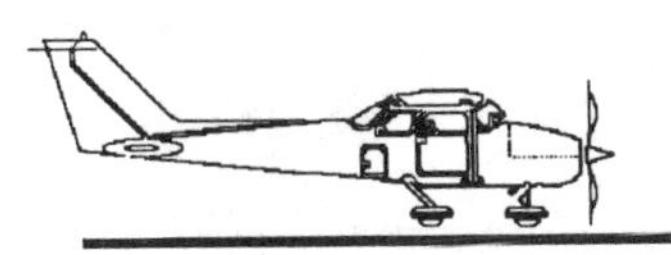
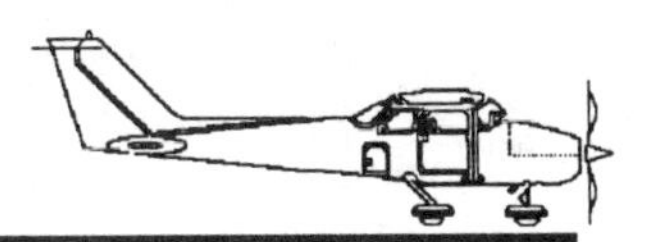

Several have intercoms installed, which is a great advantage for pilot and copilot.

Top overhauls are meaningless to the buyer. Sometimes they are done in lieu of a major overhaul, but they do nothing to avoid the need for a "major" at the manufacturer's recommended time.

More impressive than the words "500 SMOH" is the phrase, "500 hours since factory remanufacture." Those are hard to find, but several were scattered among the 300 ads. If the owner spends $7,200 with Mattituck, or $8,500 for all the trimmings (new fuel lines, new cable harness, etc.), he is not going to sell it to you a few hundred flying hours later. He wants his value out of the investment first.

A little thought is needed when looking at an ad for a 1976 Cessna 172, admittedly a good-engine year and an excellent buy, when the ad says, "200 hours since new." Where has it been all this time? In one case, it had been in an owner's barn after he became ill. When airplanes sit, they rust inside the engine. Fuel lines dry and crack, and tires go bad. Insects and birds claim the aircraft as their permanent condo, and leave secrets deep inside the airframe for the new owner to discover.

The controversy over auto gas continues with lots of emotion and few facts. Larry Miller, engine shop manager for Alphin Aircraft, Hagerstown, Md., said he has seen a lot of auto-gas-burning engines torn down for routine inspection, and they look no worse than engines that drink only avgas. (However, he declined to offer an opinion about whether to use auto gas, saying it is against policy to either support or oppose it.) You will find a lot of the Cessna 172s for sale have been converted to auto gas.

What seems the best value for the money? This analysis would indicate either a 1981 — the newest airplane on which maximum depreciation has taken place — or a 1976 model are first choices. Of course, the newer the better. As you can see, this recommendation skips those years between 1976 and 1981 when the "H" engine was used. It is probably unfair, since so many have flown their full service life with no problems. But there are just too many unknowns about the 1977 to 1980 engines.

172 Operating And Insurance Costs

If you want expert advice on Cessna 172 operating costs, ask the guy with over a decade of experience.

George A. Simmons of Herndon, Va., has operated a flying club with Cessna 172s in the Washington, D.C., area for 15 years. Over the years a Cherokee Six has come and gone, but the 172 has remained the workhorse.

Currently his club, the Herndon Sterling Fliers, has two 172s: a 1975 model is based at Manassas, Va., while the 1974 model is at Leesburg, Va. A lot of the costs presented here are club costs, such as the insurance rate. For a better example of what insurance might cost you, see the chart in this chapter.

Club Costs Are $4,000 To $4,200 Per Plane

Basically, the club operation, per airplane, costs $4,000-$4,200 per airplane, per year. For that, the club gets 150 hours of flying. Before fuel, that works out to about $28 per hour. But if you buy as an individual, or a small group of owners, your costs will be less. For example, the club needs commercial insurance, but your premium (see premium tables) if you have just 500 hours

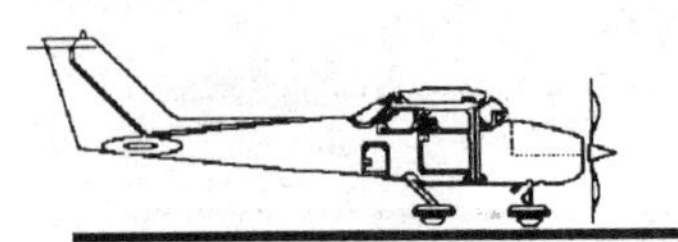

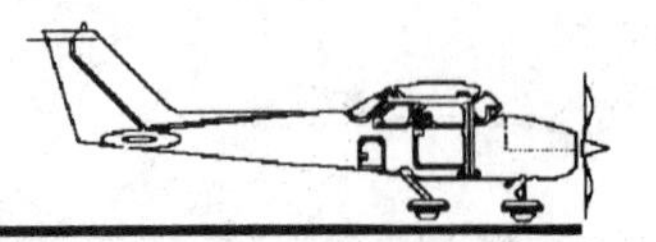

will be $980 less! And you won't need the $465-per-year answering service to schedule the plane.

Bottom line: the club operates a Cessna 172 for $42 per hour, including fuel, if they put 150 hours on it a year. **You, as an individual owner or small group of owners, could operate it for $32.** If you fly more, it gets cheaper. For example, if you double your flying to 300 hours in one year the cost of operation will drop from $32 to under $25 an hour, despite the doubled fuel bill. The reality, though, of expecting to fly 300 hours a year has to be examined. That's a six-hour trip every weekend, if you are a weekend pilot.

Simmons charges $300 to join the club, and $35 an hour for a Cessna 172 wet (less than the operating cost), and covers the deficit with a $45 monthly fee.

Simmons had especially valuable information on operating a 172 because, in the last three years, he has had both engines overhauled, new seat covers installed and both aircraft painted. The costs we show here, however, are for one airplane only.

To get the lowest costs for all the varied services an airplane requires, Simmons searches over four or five cities for the best costs, job by job.

A Breakdown Of Expenses

Here are some of the costs experienced by the Herndon Sterling Fliers.

— Insurance from a Frederick, Md., insurance firm, based on 10 club members per airplane, is $1,700 per airplane per year.

— Tiedown ranges from $55 to $85 per month, or $660 to $1,020 per year. At Leesburg you can get a "condo" hanger for an $18,000 purchase price. "I don't even go near the hangers," Simmons said.

— He uses an answering service for scheduling the plane, and that costs $465 per year.

— Routine, unexpected maintenance per year runs about $205 per airplane in little surprises. He uses, for maintenance, local repair shops like Alphin at Hagerstown, Md., (see interview with Alphin in the chapter, "The Infamous H"), Presidential Airways at Manassas, Va., and Janelle Aviation at Leesburg, Va. In other words, these are not the cheap shops or the neighbor who happens to be a mechanic.

— Tax preparation runs about $150 per year for both airplanes. (The author owned an airplane that was in leaseback to a flying school, and professional tax preparation was a necessity. You may find it helpful, even as a private owner.)

— Virginia, where both airplanes are kept, requires a $5-per-airplane license.

— Personal property tax charged by the county in which the airplane is hangered, once $300, has been revamped and now Virginia pilots enjoy a tax of only $15 per aircraft.

— There is no state income tax, since the aircraft usually operate at a slight loss each year.

— Annual inspections vary from $500 per year, to $1,500 once in three years. The reason for the once-in-three-years whopper of a bill is the "saving up" of minor, non-safety related items. For this book, Simmons used an average $1,000 cost per annual inspection. Once an annual inspection discovered a problem with the muffler — an $800 problem.

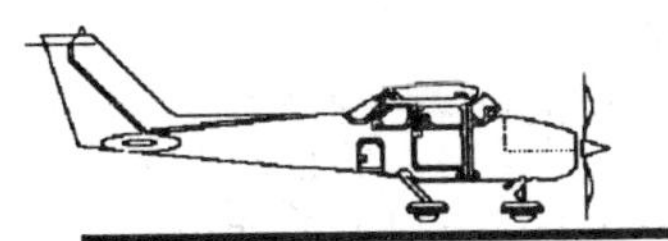
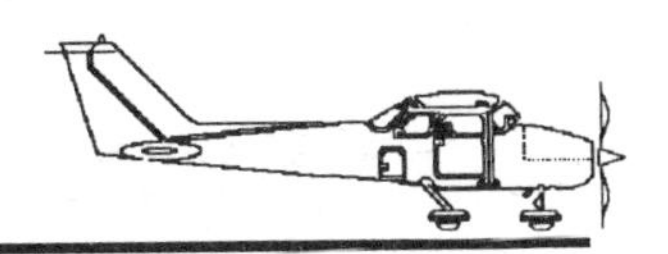

— When Cessna discovered passenger seats were sliding backward, causing an airworthiness directive to be issued, the new seat rail was $150 (round holes, not oval ones, please).

— How about something as simple as an oil change? That will be $65 for eight quarts and a new filter.

Aviation's Little Surprises

Now, for a list of the types of things that can go wrong with an airplane, including acts of God. These unexpected items were experienced by the Herndon Sterling Fliers.

— Gas fumes were detected from a cracked, rubber fuel tank vent line. Cost to repair: $280.

— The autopilot broke. Cost to repair: $170.

— A radio's output transistor went out, and it cost $96 to repair. An avionics repair shop at one of the airports incorrectly diagnosed the problem as a $300 frequency synthesizer. But Simmons didn't think the diagnosis matched the symptoms and went to another radio doctor for a second opinion. The result was a $200 savings.

— Want to do something simple like replace the air filter? That will be $73, please. The filter by itself costs $50. That is the same dollar amount as the replacement bulb for the tail beacon, by the way.

— Navigation light out on a wingtip? Well, a quick $37 ought to replace it.

— An alternator went out, and cost $350 to fix at a local repair shop.

— Now for that act of God. A small, isolated tornado found the exact spot where the Manassas, Va. plane was tied down and bashed a plastic wingtip on the ground. It cost $350 to replace the wingtip and inspect the wing. (Other planes were torn loose from their rope tiedowns and totally destroyed, some of them just 50 feet from the 172 owned by the Herndon Sterling Fliers.)

— A magneto had to be replaced, and cost $350.

— Those little plastic caps that go on the end of the horizontal stabilizer and elevator cost $160 for the whole tail, including labor.

— A nice repaint job costs $4,000. Most paint jobs are in the $3,500 range. You might find one for $3,000, but it takes a lot of shopping around and time.

— For engine overhauls, Simmons went to Lawrenceville, Va., — clear at the bottom of the state and a four-hour drive by car — and found a price of $5,000 for the no-frills overhaul. There were no accessories like wire harness and fuel lines included in the price.

— After 15 years, Simmons has learned a few tricks. For seat recovering, he went to an auto upholstery shop and replaced just the fabric insert portions of the seat cover. The plastic portions were reused. With savings like that, he was able to afford the toughest, best fabric possible. Now, two years later, the seats still show no signs of wear.

The radios, despite what you hear about factory issue radios, have not really been much of a problem. Simmons said Cessna radios should work well if they were made in 1970 or later. The older radios with tubes were headed for the repair shop every four to six months.

Next year, Simmons expects to replace the

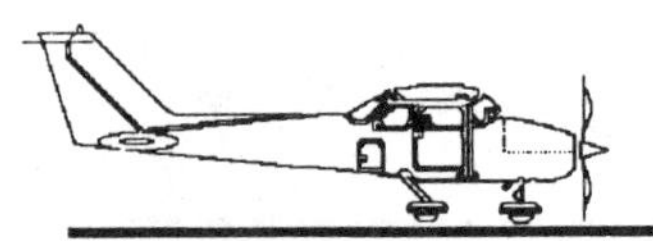
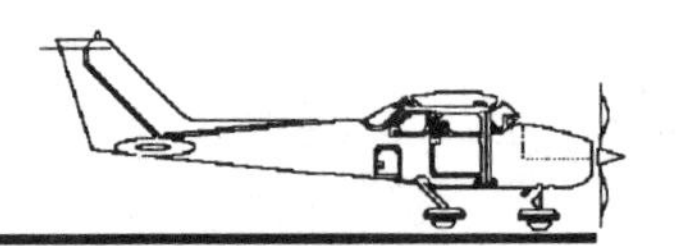

radios in one of the aircraft. The old ones will be removed and new ones put in their place, with no new rewiring. The anticipated cost of that is $1,700, he said. He believes, by the way, that 720-channel radios and mode C transponders are a necessity, especially operating in the Washington, D.C., area where there is one TCA now with the possibility of Dulles International Airport becoming a second. Leesburg lies just to the north of Dulles, while Manassas lies just to the south.

A Few Tricks Will Help

Simmons relates a few tricks for saving money, like buying recapped tires 10 at a time and sparkplugs eight at a time. Those recapped tires have, first of all, never failed, and cost only $25. A fixed base operator has to charge about $60 per tire new.

If all your maintenance is done by the local fixed base operator, you should expect labor costs in the repair shop to be $35 an hour, while those at the avionics shop are $40.

"A lot of shops start the clock running when they go out on the ramp to tow the airplane," Simmons warns. "They then charge for getting it into the hanger and taking off the cowling."

You may have heard about assisting with the annual inspection and getting a break on the price. That "break" can be as little as $50, but still it is a savings. When Simmons assisted, he was asked to do such things as unscrew screws on inspection plates and replace them when the mechanic was finished.

Simmons suspects the reason aircraft parts are so expensive is lack of automation at the plant. The manufacturer will wait for orders to accumulate, then make a single production run, which can result in delays. However, he reports availability good and gets parts in a week to 10 days. But Cessna, take heed. Simmons said if the company ever decides to stop supporting its fleet with parts, he will sell the planes and say goodbye to general aviation.

While avionics have been revolutionized over the last 30 years, the airframe has continued to be assembled almost by hand — the way it was always done, Simmons said.

Simmons has a discouraging word for those of us considering buying an airplane: unless you are a professional making a professional salary — like a doctor or lawyer — you may find it too much of a burden.

What Are Insurance Costs?

Insurance costs are made up by fortune tellers just after they have drunk a potion of pigs' tails, rat livers and bat grease. In other words, it's a science.

To get the worst rate, you would have to buy the most expensive airplane you can afford while a student pilot, then use it as a club plane. Club planes require a commercial level of insurance, which is roughly double to triple the rate given private owners. There can be up to three people listed on a private policy with no additional charges. They do not all have to be owners.

The rate you get results from adding your total time to your time in the make and model. If the result equals 1,300 hrs., you get the lowest rate. For example, the author of this book has 937 hrs. total time and 400 hrs. in a Cessna 172, so the total as far as the insurance company is concerned is 1,337 hrs.

— Rates go down roughly 10% for every 250 hours of flying experience you have. This has

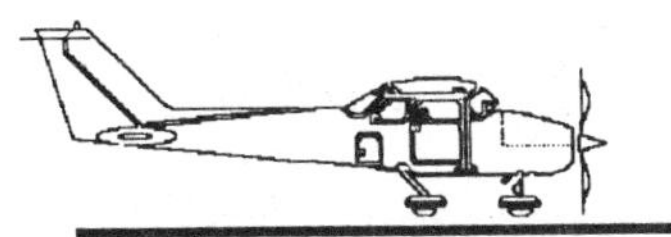
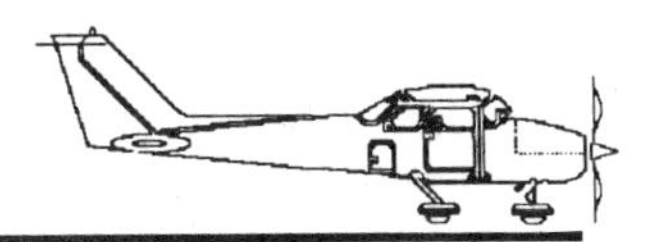

nothing to do with adding total time and time in make and model. That is figured in the back room over a boiling cauldron.

— Get an instrument rating, and your rates go down another 10%.

— If you want more than three people on the policy, the premium goes UP 10% for the fourth and fifth, thus neatly wiping out any advantage you just got for having that instrument rating.

— If more than one person is listed on a policy, it is written with a premium based on the least experienced person, just as though the other pilots weren't there.

— Most of the insurance companies have exactly the same rules, and differ slightly in premiums and corporate personality.

— The difference in premiums between student pilots and those who meet the magic 1,300 hrs. requirement is about 41%.

Let's say you buy a 1981 Skyhawk and have 1,300 hours when total time and make and model time are added together by the insurance people. One insurance company said, in March, 1988, the rate will be $760 if you paid about $29,500 for the aircraft. That will get you the standard liability coverage of $500,000 and $50,000 per person.

But if you add two additional pilots, either one a student, the rate goes up to $1,256 a year—$496 more — because of the student.

The chart below is from a midwestern insurance firm, and rates are current. The firm asked they not be tied directly to these costs, since the premiums change so often.

In the chart you see premiums similar to those given in the previous example, although they are from a different company. The rate for the experienced pilot would be $762, while the student premium is $1,160 for the same coverage.

Sample Yearly Insurance Rates For A Cessna 172*

Year	Value	Student	Pvt. 250 hr.	Pvt. 500 hr.	Pvt. 1,300 hr.
1960	$12,000	$940	$797	$733	$625
1965	13,000	965	816	751	641
1970	15,750	1,038	878	807	686
1975	21,500	1,089	920	846	718
1980	29,500	1,160	979	898	762
1982	38,500	1,289	1,085	995	842
1984	49,000	1,379	1,160	1,060	898
1986	60,000	1,433	1,204	1,103	931

* The liability limits are based on $500,000 combined single limit (includes bodily injury to the public, property damage liability and passenger liability limited to $50,000 each person).

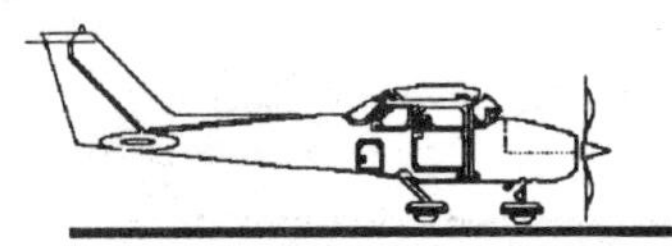
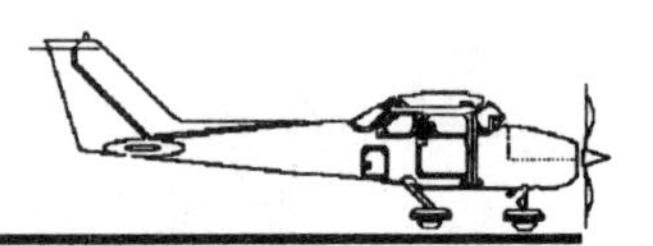

A Dealer's Advice: Van Bortel Aircraft

Van Bortel Aircraft, Rochester, N.Y., specializes in nothing but Cessna 172s and lists the phone number as 1-800-Skyhawk, although it has established a hotline for Cessna 182 sales (you guessed it, 1-800-Skylane). (See other advice from Van Bortel in the section on "The Infamous "H" Engine.")

Having previously been warned that aircraft dealers are "just like car dealers," our reporter braced for the worst: razzle dazzle lingo, self promotion and a quick brushoff. But George Van Bortel, sales manager and brother of the owner, was shockingly candid.

First, Van Bortel was asked about Federal Aviation Administration airworthiness inspector Bill O'Brien's suggestion that the buyer offer to pay for a 100-hour inspection, if the seller will pay for all repairs.

Bring On The Mechanics!

"Yes! Absolutely! I would do that," Van Bortel said. "If there is even one thing wrong with the airplane, I pay for it.

"You gotta check the airplane out. We deliver the airplane to the buyer's mechanic, not the buyer. The buyer can have the mechanic take three hours or three days going through it, and we don't even ask for a deposit. I pay for repair parts.

"If, after the inspection, the buyer calls and says the plane is not what he was promised, then I fly it back. I did that once with a buyer in New Mexico," Van Bortel said. He also emphasized, as the FAA does, that a thorough checkout of the plane's logbooks is needed. Usually it is best done by a mechanic, but the buyer can do a preliminary check to see if it is worth the trouble and expense of hiring a mechanic.

He was asked for a dealer's tips on how to get the best deal (in other words, the lowest price) from a dealer.

Price Matters Only In The Short Run

"What you should be more concerned about is getting the best airplane. There is a reason why prices are different. I have bought hundreds of airplanes, and every time I thought I stole one [at an extremely advantageous price] I was wrong," he said.

His point was that the old saying, "You get what you pay for," is more true about airplanes than other items you buy. If the airplane has a cheap price, it probably also has an expensive repair bill — after the sale.

"If you buy a more expensive airplane," Van Bortel said, "you get your money's worth, and you are more likely to get your money out of it when you sell it.

"Bargain hunting is nothing but a headache," he said. The headache occurs after you buy the airplane and head for the repair shop, of course.

Brokers Don't Own Their Airplanes

Van Bortel also cautions buyers to consider the difference between dealers and brokers. Dealers like himself own their airplanes, have looked at them and meticulously checked them over, since they were formerly the buyer of the airplane.

Brokers, while they may have access to a greater listing of aircraft in their computer than the dealer has on the ramp, often know less about each airplane. The aircraft may be hundreds or thousands of miles away.

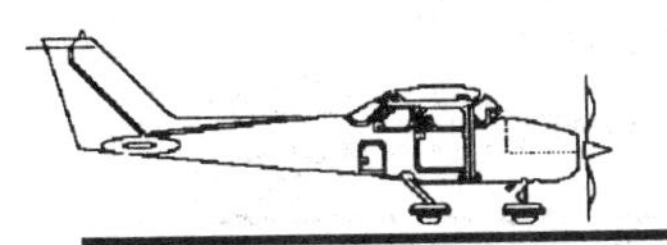
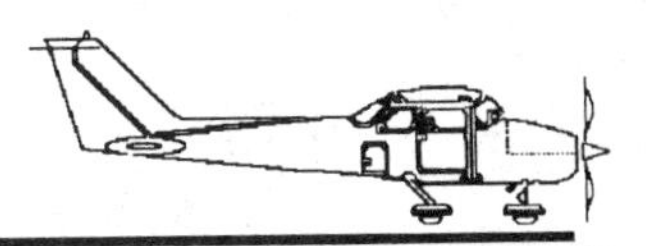

Van Bortel, while cautious, admits he sometimes buys from brokers, but hasn't always been satisfied with the result.

"Every time, the information you get is not what the airplane actually is," he said.

Van Bortel said that, despite publicity about the decline of general aviation, "Sales are as brisk as they have ever been. I hope general aviation is not in a decline. I want to think positive and hope Cessna will start up [production] again."

The same week Van Bortel was interviewed, Cessna cut back further than ever in its work force, slowing Caravan and Citation production. But there were also heard, that same week, rumors of future plans to restart Cessna's piston production line.

The Infamous "H" Engine

The Lycoming 0-320-H2AD engine is on model years 1977, 1978, 1979 and 1980. Even today most people are confused over whether it is a good engine and what years were affected.

To get answers, we called two good mechanics. The first was Larry Miller, engine shop manager of Alphin Aircraft, Hagerstown, Md. He was called because many of the aircraft owners in the Washington, D.C., area rate Alphin as their first choice.

"It's got a problem," Miller said when asked about the "H". "It eats up tappets and camshafts. Not enough lubrication gets up in there, but basically it is a good little engine."

"From a mechanic's point of view it is a good engine," he continued. It would be good, if only owners would follow directions exactly for keeping an oil additive in the engine.

Many of the "H" engines have had what is called a "T" modification, but an "H" engine plus a "T" mod still doesn't spell success. The "T" mod widened areas on the camshaft where oil can be spread for lubrication. Contrary to what some owners believe, Miller said the mod does not reduce engine power.

"The 'T' mod doesn't do the trick," Miller said.

The problem can indeed be eliminated through use of an oil additive.

"But owners forget to use it every time they add oil, and therefore it gets diluted down too much," Miller said. In other words, owners use the additive with an oil change, but when the oil gets a quart low they put in a quart and forget the additive. Do that a few times, and the additive is diluted to the point where it is no longer effective.

How can you know what the owner of the plane did with the oil additive? Ask a leading question, such as, "I hear it is a real pain to put additive in every time you add a quart of oil." If the owner doesn't know what you are talking about, he didn't put in the additive. But if he instead gives you a lecture on how important the additive is, you can be assured he uses it. (Or if he uses Aeroshell W 15W50, you've got no problems. The additive is in there.)

The second call was to an official of Mattituck, the engine overhaul company in Mattituck, New York. Their price for a factory spec overhaul for most Cessna 172 engines is $7,200. But for the "H" engine it is $8,200. 'Nuf said?

Miller said an engine — any engine, not just

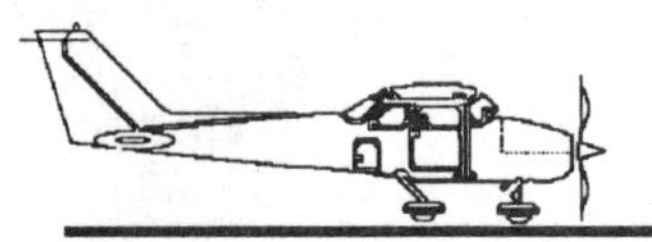

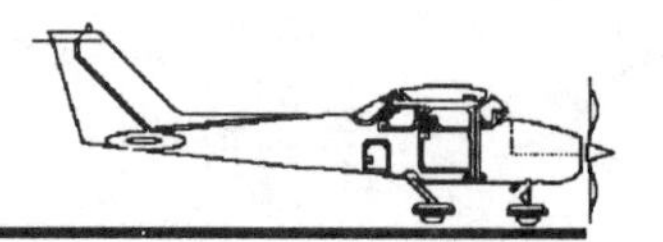

the "H" model — can be overhauled any number of times, as long as it can be brought back to factory tolerances. He was asked about stress buildup.

"Metal stabilizes over the years," he said. However, the crankshaft may have to be reworked, he noted.

Alphin does a rock bottom basic overhaul for $5,500. Most overhauls take 14 hours, but can go as high as 17. But if you want new wire harness, engine mounts, fuel and oil lines and other accessories, the overhaul will go as high as $8,500.

Miller was told about a general aviation newsletter article that rated the newer Cessna 172 engines — those since 1981 — as "fair." He disagreed, saying the rating ought to be better than that.

"The new engines are good. Most of the problems stem from the owners and operators," Miller said. He cited failure to change the oil frequently enough as one of the top engine mistreatment problems created by owners and operators.

More Advice From A Dealer

George Van Bortel, sales manager of Van Bortel Aircraft in Rochester, N.Y. — which specializes in selling Cessna 172s — has some advice for buyers of the "H" engine. Spend more on the pre-buy inspection — like $200 on the engine alone — to check that engine out in every nook and cranny. If you want an "H" you will spend more money and take more care, but there are safe ones out there to buy.

If you find a good one and buy it, take care of it. For the "H" engine, change the oil every 25 hours, Van Bortel says. "Change the oil religiously, using the additive."

Preheat the engine, expecially the "H" engine, anytime it gets below 35 degrees F., he said. He noted some owners preheat any time it is below 38 degrees, but he feels 35 is an acceptable temperature. The important thing to remember is that no engine likes the shock of waking up on a cold morning, but it is especially rough on the "H" engine.

Every 100 hours, pull the tappets out of the engine and see if they are distorted or worn. Determine whether there are metal flakes in the oil. That may be a little more expensive than the average 100-hr. inspection, but Van Bortel said it is also necessary with the "H" model.

The Lycoming 0-320-H2AD can work as well as any Cessna 172 engine, as long as the previous owner took care of it and you do the same.

The Type Certificate

Part of the buyer's job is to get down to the local Federal Aviation Administration District Office and read the type certificates on the airplane you intend to buy. (We did that for you.) That way, you be able to (1) see if the propeller, engine and other parts that are supposed to be on the airplane are really the ones you are getting, and (2) see if there are any "poison pills" hidden in the "notes" section. The notes can, but not always, include data that would require the airplane to be essentially thrown away in 11 years.

The notes section of the Cessna 172 certificate is very tame, with no nasty surprises. Most of the notes pertain to placards that must be in the airplane. If your airplane doesn't have the placards, it is relatively easy and inexpensive to get them.

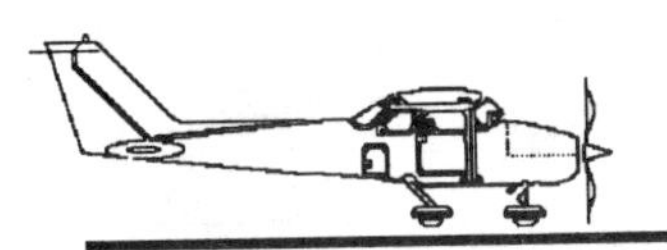

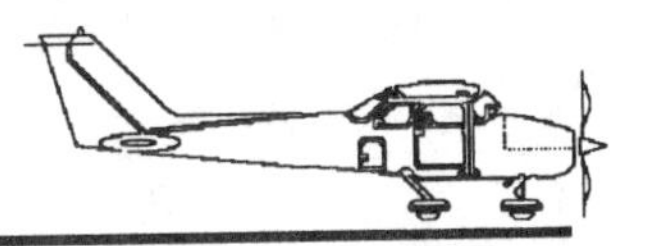

If you read one of these yourself, it is important to know the certificate for the new model is usually filed the spring previous to the model year. Thus, the 1981 Cessna type certificate was filed with the FAA in the spring of 1980.

Below is a summary of some of the data from the 172 type certificate. If your airplane does not have the right engine or right propeller, then there must be a supplemental type certificate (STC) filed for permission to change it from what you see here. Ask the owner to show them to you. If he can't, the plane is illegal.

CESSNA 172 (starting model years 1956-57). (145 hp.)

Engine: Continental 0-300-A or 0-300-B. Fuel: 80/87 minimum grade aviation gasoline. Engine limits for all operations, 2,700 r.p.m. Propeller: McCauley 1A170, Sensenich N74DR, or McCauley 1C172 MDN. Maximum weight: 2,200 lb. Fuel capacity: 42 gal. total, 37 gal. usable.

CESSNA 172A (starting model year 1960), and 172B (model year 1961). (145 hp.)

Engine: Continental 0-300-C or 0-300-D. Fuel: 80/87 minimum grade aviation gasoline. Engine limits for all operations: 2,700 r.p.m. Propellers: McCauley 1C172/EM, (seaplane) McCauley 1A175/SFC, and Sensenich M74DC. Maximum weight: 2,200 lb. Fuel capacity: 42 gal. total, 37 usable. (37 gal. usable, 172B)

CESSNA 172C (starting model year 1962). (145 h.p.)

Engine: Continental 0-300-C or 0-300-D. Fuel: 80/87 minimum grade aviation gasoline. Engine limits for all operations: 2,700 r.p.m. Propellers: McCauley 1C172/EM, (seaplane) McCauley 1A175/SFC, Sensenich M74DC. Maximum weight, 2,250 lb. **Fuel capacity: 39 gal. total, 36 usable.**

CESSNA 172D (starting model year 1963), 172E (starting model year 1964), 172F (starting model year 1965), 172G (starting model year 1966), 172H (starting model year 1967). (145 hp.)

Engine: Continental 0-300-C or 0-300-D. Fuel: 80/87 minimum grade aviation gasoline. Engine limits for all operations: 2,700 r.p.m. Propellers: McCauley 1C172EM, (seaplane) McCauley 1A172/SFC. (No Sensenich listed in type certificate for these years.) **Maximum weight, 2,300 lb.** Fuel capacity: 39 gal. total, 36 usable.

CESSNA 172I (starting model year 1968), 172K (starting model year 1969). (150 hp.)

Engine: Lycoming 0-320-E2D. Fuel: 80/87 minimum grade aviation gasoline. Engine limits for all operations: 2,700 r.p.m. Propellers: McCauley 1C175/MTM, (seaplane) McCauley 1A175/ATM, McCauley 1C160/CTM, McCauley 1C160/DTM, (seaplane) McCauley 1A175/ETM. Maximum weight: 2,300 lb. Fuel capacity: 42 gal. total, 38 gal. usable.

CESSNA 172L (starting model year 1971). (150 hp.)

Engine: Lycoming 0-320-E2D. Fuel: 80/87 minimum grade aviation gasoline. Engine limits for all operations: 2,700 r.p.m. Propellers: McCauley 1C172/MTM, (seaplane) McCauley 1A175/ATM, McCauley 1C160/CTM, McCauley 1C160/DTM, (seaplane) McCauley 1A175/ETM. Maximum weight: 2,300 lb. **Fuel capacity: 42 gal. total, 38 usable.**

CESSNA 172M (starting model year 1973). (150 hp.)

Engine: Lycoming 0-320-E2D. Fuel: 80/87minimum grade aviation gasoline. Engine limits for all operations: 2,700 r.p.m. Propellers: McCauley 1C160/CTM, McCauley 1C160/DTM, (seaplane) McCauley 1A175/ATM, (seaplane) McCauley 1A175,ETM. Maximum weight: 2,300 lb. Fuel capacity: 42 gal. total, 38 usable.

CESSNA 172N (starting model year 1977). (160 hp.)

Engine: Lycoming 0-320-H2AD (the problem engine). Fuel 100/130 minimum grade aviation gasoline for serial numbers 172261578, 172269310 through 172274009. Engine limits for all operations: 2,700 r.p.m. Propellers: McCauley 1C160/DTM, (seaplane) McCauley 1A175/ETM. Maximum weight: 2,300 lb. **Fuel capacity: 43 gal. total, 40 gal. usable.**

CESSNA 172P (starting model year 1981). (160 hp.)

Engine: Lycoming 0-320-D2J. Fuel: 100LL/100 minimum grade aviation gasoline. Engine limits for all operations: 2,700 r.p.m. Propellers: McCauley 1C160/DTM, (floatplane) McCauley 1A175/ETM. **Maximum weight: 2,400 lb.** Fuel capacity: 43 gal.total, 40 gal. usable.

Nearly all the "notes" contain information about placards that must be on the airplane to describe limitations and operating procedures for the aircraft. A mechanic can assure that the placards are in compliance with the type certificate.

Cessna 172 Service Bulletins And ADs

You'll have to let a mechanic check to make sure all the Airworthiness Directives (ADs) have been complied with. But here is a sample of the kind of expensive or inexpensive repair orders they can contain.

Old ADs Still Have a Bite

The ADs you have to fear, or rather the ones your bank account must fear, are those that are "recurring." No matter when they were issued, they must be accomplished at certain intervals. So, depending on the model year you get, it may come to you with a built-in maintenance bill over and above routine maintenance. It's a cruel world.

How would you like to get a notice in the mail requiring crankshaft replacement on specific engine serial numbers? That is what happened on Sept. 12, 1978, to Cessna 172 owners with serial numbers 17,267,585 through 17,271,034. Earlier that year owners had been hit with ADs requiring impulse coupling inspections on all Bendix magnetos, and an oil pump driving impeller replacement on certain serial numbers (again, 17,267,585 through 17,270,492).

In the next year, there was an AD about: the Bendix rotor housing interference, cigar lighter wiring, checking torque on engine mounting bolts (a relatively cheap problem), and — on Oct. 14, 1979 — vented fuel caps (another cheap problem).

A Partial List

Here's are the more interesting — not the complete list — ADs up to the present to give you an idea of what to expect:

— March 4, 1980, "H" series engine lifter replacement and oil additive;

— July 14, 1980, "H" series valve spring seat identification;

— Sept. 16, 1981, inspection of elevator cable clevises;

— April 18, 1981, replacement of sintered iron oil pump impellers;

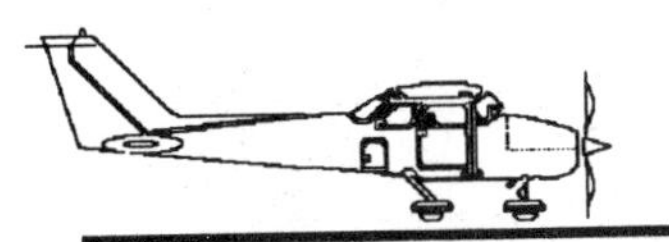

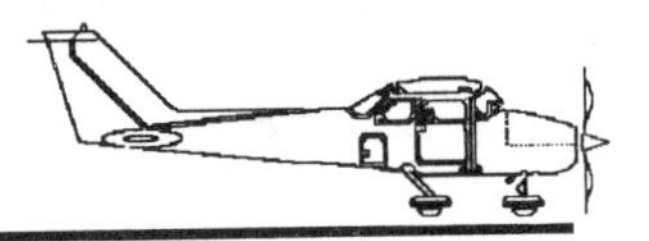

— Feb. 7, 1982, aircraft with 180 hp Lycoming, installed by STC, modify crankcase breather;

— June 17, 1983, aileron balance weight check on Robertson modified aircraft;

— June 22, 1983, aileron hinge inspection;

— July 23, 1985, King 100 and 150 autopilot systems. Purpose: to prevent undetected trim malfunctions that may result in significant mis-trim conditions.

— Oct. 14. 1985, Hartzell Propeller Products Division, to prevent propeller blade clamp failure.

Obviously, the March 4, 1980 AD is of special historical note, since it served to confirm the continuation of problems with the "H" engine used in 1977, 1978, 1979 and 1980. It is in effect today. Now, however, Aeroshell Oil W 15W-50 has the additive in it. As long as the owner uses that oil, the oil additive AD is complied with.

The Newest ADs

Here is a sampling of the most recent ADs for the Cessna 172:

— Feb. 5, 1986, United Instruments altimeter AD to prevent possible erroneous altitude information from being displayed to the pilot;

— April 26, 1986, to prevent slippage of the pilot/co-pilot shoulder harness;

— Oct. 22, 1986, Collins Avionics Division, applies to the Collins model DME-42 distance measuring equipment, to prevent display of erroneous information;

— Nov. 19, 1986, Cessna, to eliminate the possibility of engine power reduction due to contaminated fuel (installation of quick drains);

— March 20, 1987, Cessna, to assure proper locking of seat mechanism and to preclude inadvertent seat slippage;

— June 10, 1987, Avco Lycoming, affecting certain parallel valve-type engines that were remanufactured between July 1, 1985 and Oct. 8, 1986, to prevent possible rocker arm failure and loss of engine power. Inspect and rework rocker arms.

Service Letters are a lot less exciting than ADs. For example, the ninth letter of 1987 was simply a new Cessna sales and service directory. Quite often they are simply informative in nature, with no maintenance bill required. Many of them cover the same topics later made into ADs.

What Will All Those Recurring ADs Cost?

The best advice about ADs is to ask the pre-purchase mechanic if he can give you some idea as to the total yearly cost of the "recurring" ADs. If there are some non-recurring ADs that have not been complied with, naturally you will ask the seller to bring the aircraft up to date prior to sale.

Cessna 172 Accident Reports

An examination of accident records gives an idea of how the Cessna 172 performs in the field, and which systems are the most prone to failure.

The most interesting discovery for Cessna 172 buyers — from a recent analysis — is the absence of the Lycoming 0-320-H2AD engine in the accident reports. When it IS there, it was

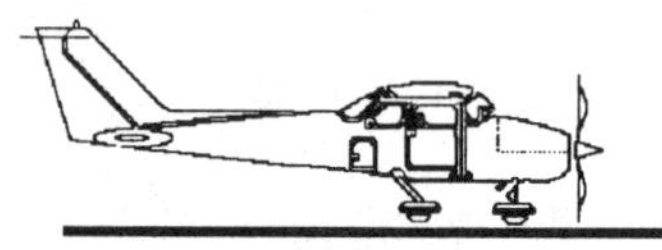
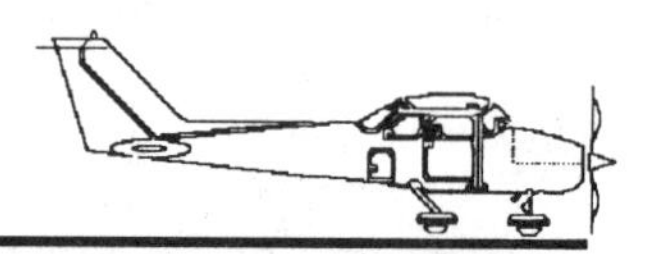

usually along for the ride while other causes created the accident.

The National Transportation Safety Board (NTSB) was asked for a listing of mechanical failures for the period 1983 to 1988. It found 195, or about 36-40 per year.

But a closer analysis of each accident in 1982 — one of two years for which NTSB provided details —reveals only 19 of the 36 accidents were actually due to mechanical failure. The rest were clearly due to other causes, usually pilot error.

In 1985, only 16 of the 40 accidents listing a mechanical failure as a cause were actually due to a mechanical problem as the "primary" cause. NTSB also threw in two 172RGs and a 172XP, but those were removed from our 172 analysis.

Under the NTSB filing system, if a pilot runs out of fuel, botches the forced landing and dives into the ground, all those parts which broke on impact get listed as "mechanical" causes. Then, when an analyst asks for "mechanical" failures, all accident reports with the slightest mention of bent metal or broken parts are picked out of the computer.

For example, in 1982 a 76-year-old pilot hit a snowbank he didn't see on an improperly plowed runway. The right main gear sheared off, and the incident popped out of NTSB files as an accident caused by a mechanical failure. But the landing gear worked fine until it hit the snowbank.

We threw out all accidents listed as "mechanical" in which the pilot ran out of fuel, steered himself into a mountain, forgot to replace the oil filler stick, left water in the fuel, hand propped the plane with an untrained person at the controls, tried to fly with frozen fuel lines or forgot to use carb heat. There were a lot of accidents like those classified "mechanical."

Don't Trust Fatality Statistics, Either

So, be suspicious of statistics about mechanical failures from NTSB. But fatalities are also used to judge an aircraft's safety record. In 1982 there were five fatalities due to accidents in which mechanical failure was listed as one of the causes. Better be suspicious of those statistics, too. Let's look at all the fatal accidents.

— A 32-year-old pilot took off from Woodruff, Wis., after receiving weather briefings that included severe weather associated with a cold front and thunderstorms. The 122-hour pilot, who had one hour of instrument time, told the ground he was "totally obscured." At that point the heading indicator of the Cessna 172C failed and he crashed into a wooded area shortly afterward.

— A 67-year-old pilot and one of two young boys strapped together in the right front seat died when an "up and down" flight path — to thrill the boys — kept the approximately one pint of fuel in the right tank from entering the engine. The fuel selector was on the right tank of the 172M.

— A 30-year-old pilot was seen prior to takeoff brushing frost off the wings with his hands, and was heard to ask for a window scraper, before departing an airport at an elevation of 9,927 ft. The aircraft was unable to gain altitude after liftoff and crashed with the left wing low.

— The final accident, claiming the fifth victim, appears to be the only one in which the airplane could be implicated. A 28-year-old student pilot with 17 hours was killed after an emergency landing in a wooded area near a pasture. Examination revealed the oil pump drive gear failed.

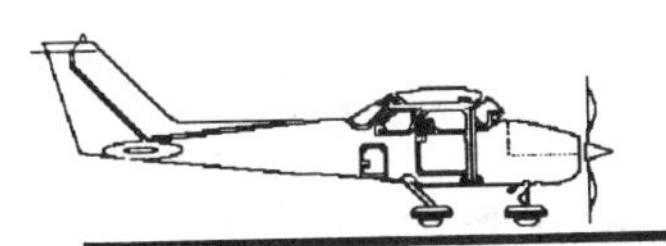
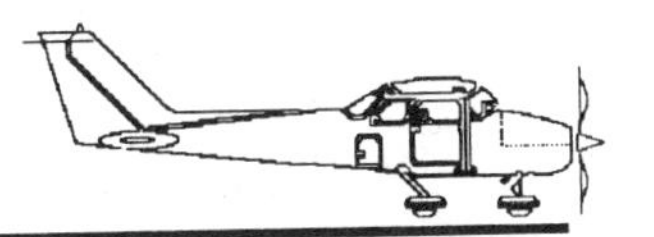

A Look At Other 1982 Mechanical Failures

Now, let's look at the remaining, "real" mechanical failures of 1982:

—Wheeling, Ill., Jan. 24, electrical fire—on the ground—under the instrument panel;

— Brownsville, Ky., Feb. 8, electrical fire with blue sparks coming from under the instrument panel;

— Thayer, Mo., March 17, pilot's seat slid full aft and pilot was unable to reach the controls;

— Salem, Ohio, April 18, carburetor float stuck open and flooded the Lycoming 0-320-2D engine;

—Houston, Tex., May 16, plane swerved on landing because the left brake pad was missing;

—Greenville, Maine, June 1, floatplane 172 suffered two partial losses of power during takeoff and settled into a wooded area;

— Martinsburg, W.Va., June 29, engine oil pump failed when a gear tooth in the pump failed and became wedged in the gear housing. An Airworthiness Directive (AD) on the pump had been issued in 1981, but the pump had not been in service long enough for the AD to apply;

—Lander, Wyo., Aug. 3, nose gear fork broke on an aircraft used for student training, which had a history of use on rough and unimproved fields;

— Cheboygan, Mich., Sept. 9, rubber foam from the air intake had dried and shredded and was ingested into the carburetor air intake, causing the Lycoming 0-320-E2D engine to stop;

— Vinalhaven, Maine, Sept. 11, mixture control cable broke with the mixture in the idle cutoff position;

—Prospect, Ky., Sept. 13, an alternator brush was completely worn, resulting in failure of the electrical system;

—Millry, Ala., Sept. 15, right exhaust flame tube separated, causing loss of power in the Continental 0-300-D engine of the Cessna 172F;

— Brimfield, Ill., Sept. 18, number four exhaust valve of a Lycoming 0-320-E2D engine failed after flying 2,156 hours without an overhaul, causing partial power loss in the engine which had oil changes only every 200 hours, instead of the recommended 50 hours;

— Lansing, Ill., Oct. 24, electrical failure caused by an unidentifiable noise filter which was installed in the line of the output side of the alternator;

— St. Ignatius, Mont., Nov. 12, passenger door opened and the distracted pilot attempted to stop on the runway in snow and ice.

What is interesting about 1982 is that all four model years of the Lycoming 0-320-H2AD engine (referred to as the "H" engine), the one that got so much publicity for being a clunker, were now flying. While model "H" engines were on some of the above aircraft, it was implicated in only one, and that was a failure of the oil pump drive gear.

In 1985 several of the Cessna 172s involved in accidents were wearing the "H" engine, but as in 1982, only one accident could be linked to the engine. In that accident, the sequence of trouble began when the pilot left the oil filler stick either

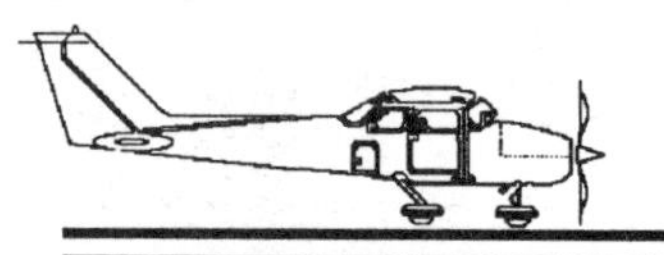
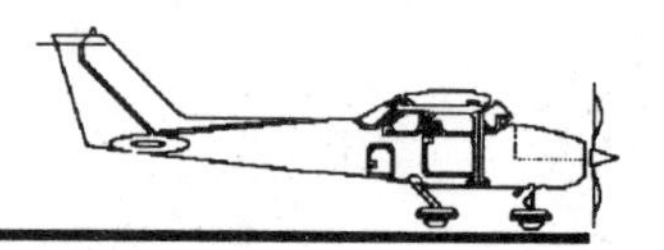

too loose or completely out after the preflight. However, there was also oil leaking from the top of the engine, NTSB said, and that saved the pilot from having to accept full responsibility. But what if the pilot had secured the oil filler stick? Even with the leak?

A second 1985 incident involving the "H" was actually due to an improper carburetor heat control linkage.

By 1985 not only were all model years of the bad engine flying, they were aging as well. Yet the results looked good for the Lycoming O-320-H2AD.

A Look At 1985 Mechanical Failures

Pilots in 1985 ran out of fuel six times, left water in the fuel twice and failed to use carb heat three times — and those accidents were classified as mechanical — but we've said enough about NTSB record keeping. Let's look at the remaining legitimate 14 mechanical failures:

— Mulino, Ore., April 7, throttle cable housing separated in the area where it is clamped to the rear of the engine mount, causing a forced landing with no injuries;

— Macedonia, Ohio, April 22, no injuries after emergency landing from cruising altitude of 2,500 ft. when the number three exhaust valve failed on the Lycoming 0-320-E2D;

— Middletown, Pa., April 27, engine crankshaft showing stress and corrosion contributed to power loss of the Lycoming 0- 320-D2G engine causing a forced landing and no injuries;

— Sharpes, Fla., April 27, engine failed during power-off stall demonstration for passengers because the carburetor had rusted parts and needed an overhaul, causing a forced landing an no injuries;

— Valley Center, Kan., July 3, plane ran off the runway because of an aborted takeoff when the controls didn't "feel" right. The elevator trim position indicator did not indicate the actual position of the trim tab;

— Leesburg, Fla., July 17, electrical system failure preventing the pilot from keying the mike to turn on runway lights, resulting in a landing on a darkened runway;

— Arcadia, La., Aug. 3, plane ran off the runway from an aborted takeoff due to partial power loss when the carburetor partially failed, no injuries;

— Sallisaw, Okla., Aug. 14, electrical failure led to a precautionary landing, but the landing was diverted to grass beside the runway when the runway became blocked by equipment, no injuries;

— Saint Helena, Calif., Aug. 14, stuck intake and exhaust valves in the number one cylinder of the Lycoming 0-320-D2J led to an emergency, partial power landing with no injuries;

— Hollywood, Fla., Sept. 24, propeller blade separated one foot from the hub due to fatigue cracking, causing an emergency landing and no injuries;

— Emmett, Idaho, Oct. 4, plane overran the runway due to worn left brake pads and leaking left brake actuator cylinder. Pilot refused to take a breath test to check for alcohol;

— Millington, Tenn., Nov. 3, Lycoming 0-320-E2D engine failed on takeoff because wires were disconnected from the magneto switch,

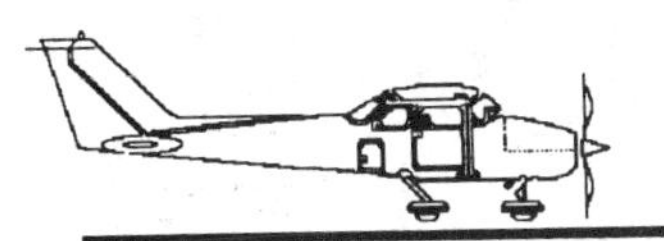
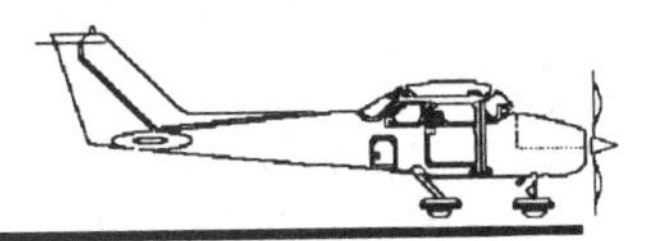

minor injuries;

— Monroe, La., Nov. 8, Lycoming 0-320-D2J failed at 5,500 ft. after failure of the accessory drive gear assembly drive gear, forcing a landing on the median of a busy highway, resulting in one minor injury;

— Bloomsburg, Pa., Nov. 28, electrical failure occurred when the alternator shorted out due to internal structure wear and the pilot attempted a precautionary landing in a muddy field, with one minor injury.

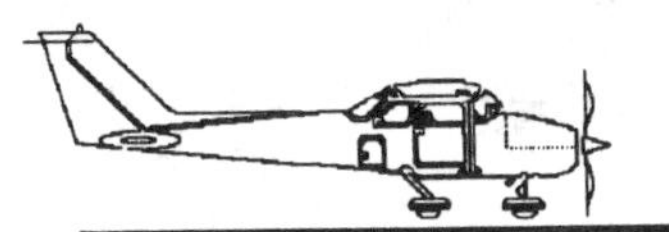
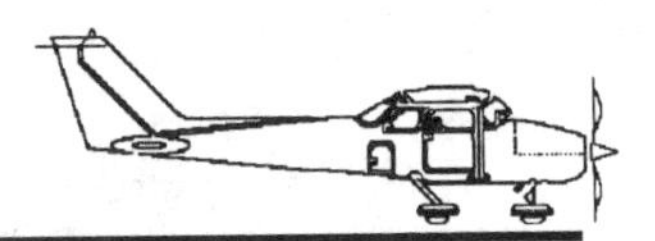

CHAPTER FOUR

THE CESSNA 150/152

How Much 150/152 Can You Afford?

This section will not repeat the generalized buying tips included in the chapter, "How Much Cessna 172 Can You Afford?" So let's get right to the 150/152.

Lead from the 100LL grade aviation fuel is a problem in every Cessna 152, but less so in the 1983, 1984 and 1985 models. That's a tip from the piston product support representative at Avco Lycoming Textron. If you look at the Cessna Serial Numbers chart (see Appendices) you notice the power of the Cessna 152 drops mysteriously from 110 hp. to 108 hp. starting in 1983.

That is because Lycoming redesigned the cylinder head that year. The new design dropped engine compression from 8.5 to one, down to 8.1 to one. All of the Cessna 152 engines from 1978 to 1982 had a severe leading problem. The lead would pool right at the sparkplug. Those engines are called the 0-235-L2C. The newer design in 1983, 1984 and 1985—called the 0-235-N2C—cured that problem by the cylinder head redesign, but the horsepower dropped by two.

"If you are trying to avoid the leading problem, the 0-235-N2C is the way to go," the Lycoming official said. He warned, however, that a pilot can defeat the reduced susceptibility to lead by idling too slowly. "The lead collector has to be kept at 800 degrees, and that means idling at 1,000 rpm. or higher," he said.

The Continental 0-200-A was used from 1959 to 1977, and operated quite well on 80 grade aviation fuel, but was caught by the leading problem like all other engines when the oil companies converted to 100LL. (Some owners joke that the LL in the name implies double lead, not low lead.)

By the way, Lycoming says the 0-235-N2C is really a 116 hp. engine. Since some of that power is used to drive accessories, the power that finally arrives at the prop is 108 hp.

So if you want to avoid lead get a 1983 or newer Cessna 152, but they start at $22,000 in the used market. A 1985 Cessna 152 advertised in *Trade-A-Plane* in Herndon, Va., recently had an asking price of $41,000! Blue book price was $2,000 less.

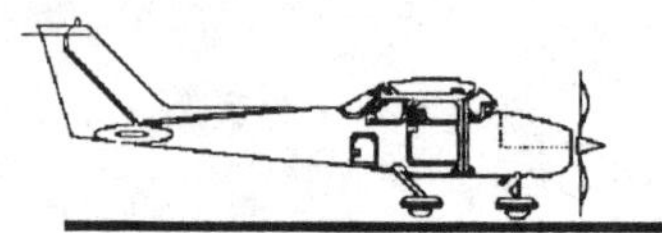

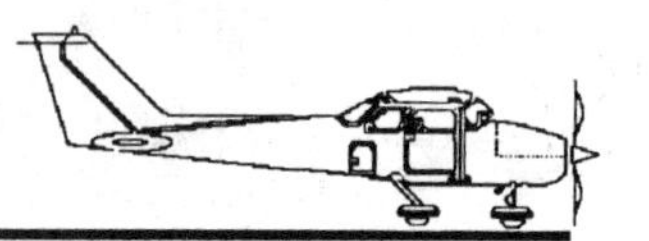

In a recent issue of *Trade-A-Plane* there were 90 150s and 85 152s advertised. Of those, only eight were 1983 or newer model years. The largest group advertised was 152s built between 1978 and 1982. Next came 30 150s built from 1974 to 1977, followed by 28 150s built from 1964 to 1968.

Dealers (see our interview with "Miss 152") think the best value for future resale is an Aerobat, whether it be a 150 or a 152. But they are scarce. Out of 175 Cessna 150/152s advertised, only one was an Aerobat 150 and three were Aerobat 152s.

There were seven Cessna 150/152s that had been converted to taildraggers, but most of those were in top shape and therefore very expensive.

We don't have to warn you that trainers have had a rougher life than most 150/152s. The problems come from not only hard landings, but also from wing flex which can rub holes in the gas tank and poor engine operation by inexperienced pilots. (How many times do you suppose a student pilot forgets to keep idle rpm. at 1,000?)

The charts below give you an idea of current asking prices.

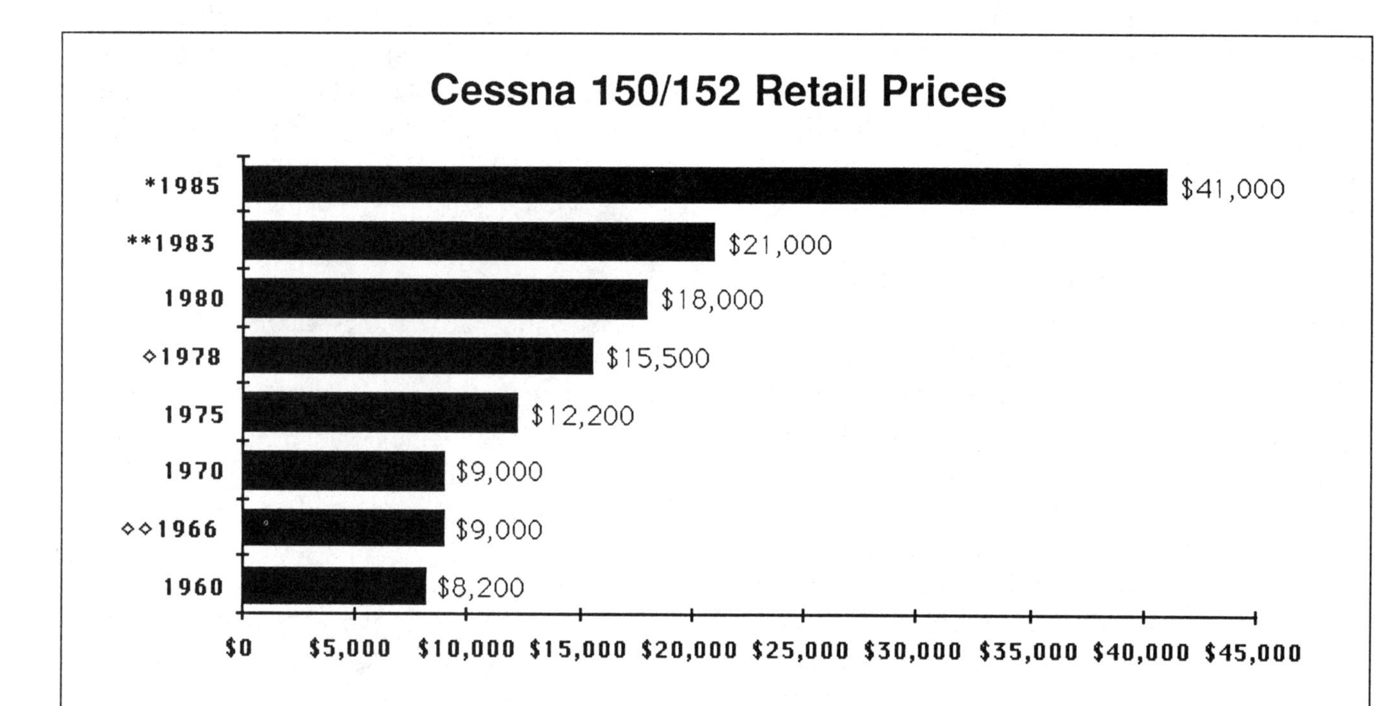

* IFR, low time, 108 hp. Last model year of 152s.

** First year of 108 hp. engine, 0-235-N2C, designed to combat leading problem.

◇ First 152, first 110 hp. engine, 1,670 max weight.

◇◇ 1,600 lb. max weight began with 1964 model.

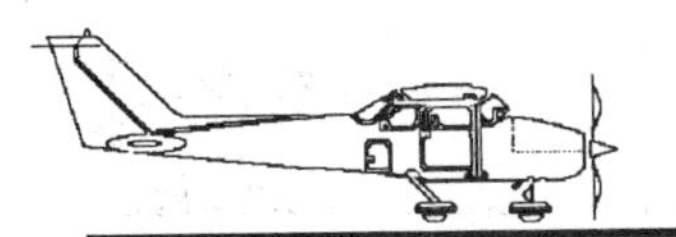
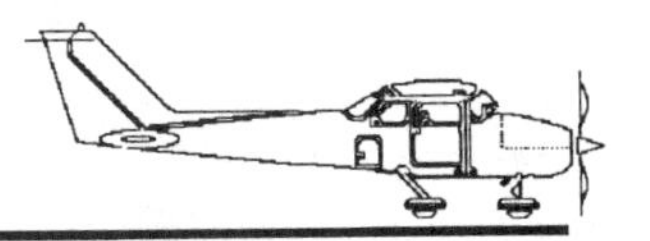

152 Operating Expenses

The Cessna 152 registered as 6159Q, a 1982 model, was purchased in October 1981 and is today flying in Ft. Lauderdale, Fla., at Executive Pilot Service. It was in leaseback the first four years after leaving the factory. Because so many 150/152 aircraft end up in leaseback, we present the following case history. For the private owner, it offers insight into what sort of expenses to expect.

First, a word about leaseback results. From November 1981 to February 1985, the plane broke even exactly 10 months. Unfortunately, that period is 40 months long. Those were looong months for the two owners, who sometimes paid $600 per month out of pocket. There were of course months when nothing was paid.

Expenses were of course higher than for a privately operated airplane. Success can cause failure in the leaseback business. Success, a large number of hours flown, means maintenance charges must be paid more frequently. The amount offered to the owners by the first FBO did not cover expenses, and there was no break given on maintenance, even though the plane was important to both the owners and the FBO. The second FBO, Center-Line Aviation, Addison, Tex., deserves much praise. First, the number of hours needed per month to break even dropped from 75 to 65 because of the excellent management of Center-Line officials and because Center-Line offered the owner almost double the return of the first FBO ($26 versus $12 per hour).

In 1982 it was calculated that some 900 hours were needed to break even. (The plane was purchased for $30,500 at a 15% interest rate.) Only 543 were flown. The switch to the second FBO occurred in late 1983, when 873 hours were needed to break even but only 704 were flown. In 1984, 792 hours were needed to break even and 808 were flown. Only one of the better-than-break-even months occurred at the first FBO.

During 1982, when 543 hours were flown, maintenance costs were $1,923, while insurance was $2,162 (most months the payment was $165 with a year-end payment of $392). The monthly payments on the aircraft were $575. There was no tiedown fee, since the aircraft was part of the rental fleet. Interest was about $3,236 per year. **A private owner would have paid $23 per hour to operate the airplane that year. That figure was calculated by reducing the insurance rates to private non-commercial levels and adding in tiedown costs.**

During 1983, when 704 hours were flown, interest was $3,236, maintenance was $4,340, insurance was $2,031, fuel was $1,093 and local and state taxes were $345. **A private owner would have paid $16 per hour that year, again adding in tiedown costs and reducing insurance to what a private pilot would pay.**

During 1984, when 808 hours were flown, insurance was $2,339, maintenance was **$5,415** and interest paid was $2,721. Fuel costs were $8,712, and property taxes were $162. **A private pilot would have paid $23 per hour to operate the airplane that year, including all costs.**

Just to get an idea of what kinds of things can go wrong with a Cessna 152, let's take a look at the major maintenance problems of 1984:

> A $14 landing light cost $9 in labor to replace.

> A new tachometer cost $45.

> An oil change and filter was $18.

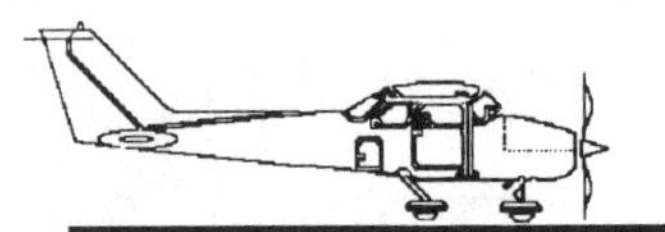
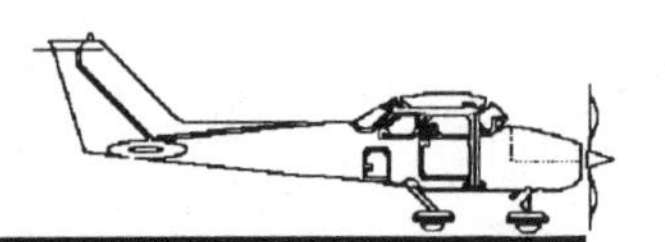

> It was $22 to wash the plane and $9 to file a nick out of the prop.

> A 100-hour inspection was $175 for labor, and the management of Center-Line provided parts they had purchased separately at discount.

> A voltage regulator cost $83 for parts and labor.

> Parts and labor for a spark plug change totaled $184.80.

> An annual inspection cost $364.36, and required two new tires, brake pads and an oil change.

> An ELT battery cost $60.

> A new alternator cost $216 for parts and labor.

> A 50-hour inspection was $58.50, and included and oil change, filter, battery serviced, brake fluid, air for all tires and strut, replacement of a worn elevator bushing, unblocking of the airspeed indicator and replacement of a panel light bulb. (You won't find one at that price.)

> Repair of the right fuel tank, cracked in several spots and leaking, $108. A few days later the airplane would not start and it cost $165 to check compression, timing, and the carburetor and install new spark plugs.

> Another 100-hour inspection cost $390, and included repair of **yet another** crack in the right fuel tank. This time it was removed and welded to stop the tank from rubbing holes in itself, and a fuel system part was replaced.

> When the nose gear shimmied about the same time an oil change was needed, the repair was $111. The nose gear shimmied again a month later, and it cost $290 to fix.

In the colder months, the 1982 Cessna 152 had several episodes of refusing to start, and this was Dallas! However, it gets cold in Dallas, as their history of ice storms and ice-cold wind off the plains to the north shows.

Despite the adversities, this history shows you save at least $15 per hour by owning a Cessna 150/152, at today's rental prices of over $40 an hour.

Sample Yearly Insurance Rates For A Cessna 150/152*

Year	Value	Student	Pvt. 250 hr.	Pvt. 500 hr.	Pvt. 1,300 hr.
1960	$6,000	$652	$556	$513	$442
1973	11,900	848	720	662	564
1978	15,500	902	765	702	597
1983	22,000	952	804	739	627
1985	39,500	1,162	978	896	759

* The liability limits are based on $500,000 combined single limit (includes bodily injury to the public, property damage liability and passenger liability limited to $50,000 each person).

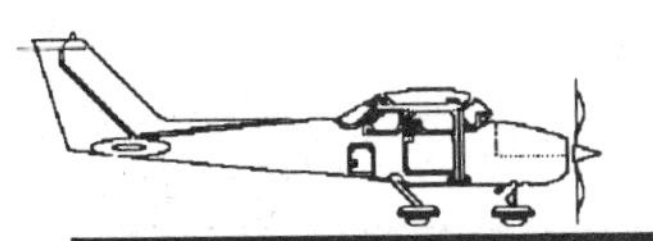
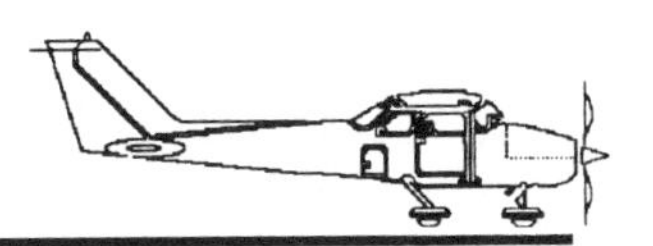

A Talk With "Miss 152"

The greatest threat of a mid-air collision today comes from two dealers trying to get to the same airport to make an offer on the same plane, jokes Vickie Scoones, the "Miss 152" at Vickie of Vermont in Burlington.

Yes, it's a sellers market right now, but that is not necessarily bad news for those of us who want to buy. Now, you can buy and fly with the confidence that when you sell you will come out way ahead, she says. However, in a market like this the buyer should decide early that the plane he buys is always up for sale. Also, the buyer must be more careful to do what Vickie calls "buy right," which means at the right price.

Weak Dollar Breaks Up the American Fleet

Vickie said it is the weak dollar in Europe that is causing much of the excitement in the market, and it is not limited to 152s. Any single piston in good repair draws interest from Europe. With the dollar weak, America has become Europe's discount toy store full of real airplanes. But there are just so many 152s. Her business has been hurt by the situation as well, because the Europeans are even outbidding U.S. dealers.

Not everyone should wipe out their savings to buy a 152 for sale to the Europeans, however. There are all kinds of complications involved with de-registering an airplane in the U.S., faxing documents to Europe and finding proper crates for shipment of the aircraft.

Vickie's advice is to do a really careful pre-purchase inspection (which you can do with this book). When she buys a plane, and she has bought dozens, she pays a mechanic $250 a day plus expenses to travel to the plane's location and check it over.

If there is something wrong with the plane, should the dealer pay?

Not Everyone Wants A 'Perfect' Airplane

"Normally, I pay. I want it to go out right. But most of our planes go out with fresh annuals. Sometimes dealers will call with a request to buy 152s for a flight school, and they want to do their own maintenance. They just want the planes at the lowest cost, in that case," the professional musician said. She is a professional flutist, and admits to playing one made by the Japanese because, "It is better and cheaper."

Some Advice From Miss 152

"A 152 is a good investment, because a year from now it will go up," Vickie said. "People never lose money." One man was especially nervous about whether the airplane would appreciate, so Vickie told him if it didn't, she would gladly buy the airplane back. (She can always sell it to Europe. Dealers in Europe are getting as much as $55,000 for a 152.) Other dealers have a similar offer, with Van Bortel offering to buy it back in the first 30 days or 30 flight hours, whichever comes first.

"Set your price range, buy the plane, then fly it for 100 hours and sell it for more," Vickie advises.

Surprisingly, most of her customers have done their research before they arrive. Those that haven't drive her almost as crazy as her telephone, which never stops. One customer found nothing wrong with the plane, but wanted $500 off.

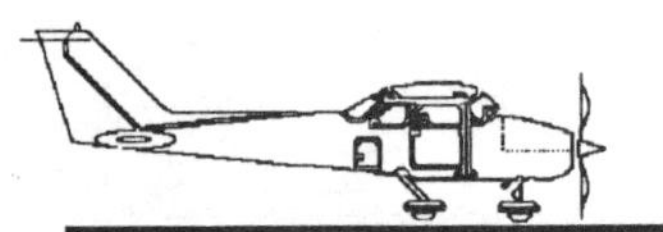
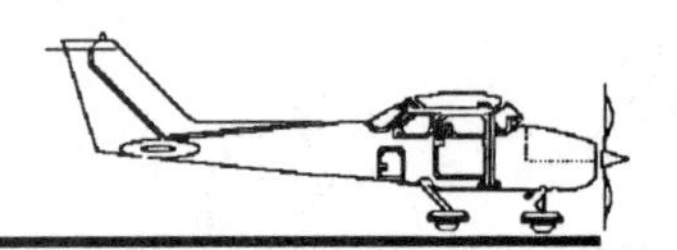

"I told him, you just wasted my morning," she said. And she walked off. But she also helps buyers with advice, she said.

For example, Vickie said the 1978 152 has a poor reputation for attempted starts in cold weather. Where she lives, they capitalize the word Cold.

The most popular year, the one customers ask for most often, is 1980, she said. Customers trust her, "Possibly because I'm a woman," she said, and come back when they are ready to trade. One customer has bought six planes from Vickie.

The quality of the plane does make a difference, in addition to the model year. One gentleman had a 1978 he bought new that was in showroom condition — never used as a trainer — with under 600 total hours on it. While many 1978 152s would sell for $11,000 due to high time, his sold for $22,000.

Now, For The Best Advice

The very best model you can buy, Vickie said, is an Aerobat. It makes no difference whether it is a 150 Aerobat or a 152 Aerobat. Obviously, you want one with low time, and 1,000 hours on the engine is considered reasonably low. Also, it must have no damage history. Thus, a few extra dollars to a title company for an accident search is worth the effort.

[A Lycoming product support official also had good advice about the "best" 152. Get a 1983 or newer model, he said, because those have a redesigned engine able to cope with high lead 100LL gasoline.]

One Aerobat with exceptionally low time sold to — you guessed it — Europe for $45,000.

The buyer looking for low prices must realize the price goes way down for the 4,000 hour airframe (but the maintenance cost may go the other way.)

Obviously the 152 is a great investment now, and there may never be another batch made. Vickie warns that the days of the more sophisticated trainer are just around the corner.

A Talk With Carter Aviation

One way to judge an engine overhaul shop, other than by reputation, is friendliness — willingness to stay with you until you understand the prices and services. Such a shop is Carter Aviation Supply, a folksy shop that prints, "Howdy Y'All" in its ads. Bob Carter preaches the truth in engine overhaul with the fervor of a Bible Belt preacher.

Carter explained not only his prices, but how he computes them and his actual costs to a caller one day recently. The caller had just been brushed off by another shop in Indiana with, "I'm busy now." Carter was a welcome change. His shop is in Elizabethton, Tenn.

He was asked for not only factory overhaul and remanufacture prices, but new engine prices as well. He said Continental gives dealers 25% off, while Lycoming gives dealers 20% off list price — the price you pay. So the dealer's profit is normally 25% for Continental and 20% for Lycoming products.

Here are some of his prices. (All prices are in addition to the "core" amount which is refunded when the trade-in engine reaches the factory). ***See chart, next page.***

If you see a "rebuilt" price, that means Carter will rebuild your own engine, rather than exchange it for a factory overhauled or remanufactured engine.

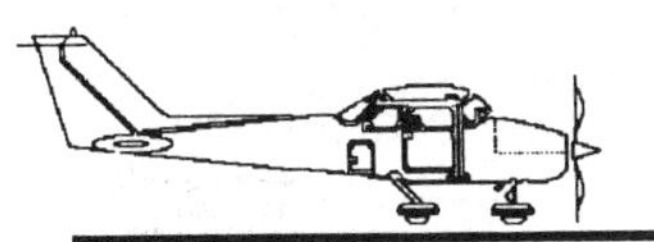

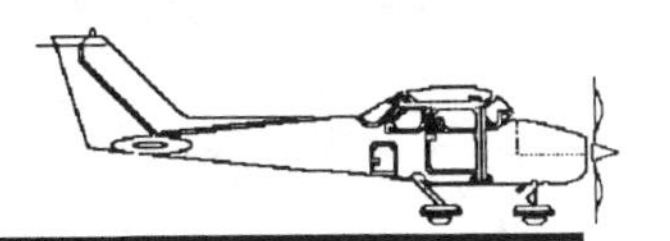

FACTORY PRICES

	Overhaul	Remanufacture	New
Cessna 152 0-235-L2C (Core is $4,400)	$6,892	$7,692	$9,000
Cessna 172 0-320-D2J (Core is $5,200)	$6,884	$9,532	$12,580
Cessna 182 0-470-R (Core is $6,600)	—	$8,383	$14,719
Cessna 210 TSIO-520-R (Core is $6000)	—	$14,431	$21,799

"You call me and I'll tell you how it is," Carter said. (1-800-251-1616). "You can get burned [financially] if you don't know what you're doing." On subsequent calls Carter held true to his word. He does indeed "tell it straight." He was asked why the core cost of a Cessna 182 engine should be higher than that of a Cessna 210 engine. Carter said the liability insurance on an engine part in the Cessna 182 engine went up 1,000% when the parts firm was sold. Details like that can keep the owner/pilot from — as Carter says — getting burned.

Some Final Tips

Here is what Carter recommends. On the Lycoming small, "O" series engines, always swap engines because you'll get a "like new" engine except for the crankcase and crankshaft. On the "IO" series, try to rebuild your own engine because you can save $3,000 and — on the more expensive engines — sometimes $10,000.

On the Continental 0-470, always swap engines, but if you have an IO-520, rebuild your own engine if possible.

The 150/152 Type Certificate

Like the 172, the type certificate for the 150/152, which uses one certificate in FAA records for both planes, contains no nasty surprises in the "notes" section. The notes pertain to the center of gravity for unusable fuel and undrainable oil, and placards that must be displayed on the various models. Here are the specs on the model you may be interested in buying, as taken from the type certificate.

CESSNA 150 (First Model Year — 1959), 150A (1961), 150B (1962), 105C (1963).

Engine: Continental 0-200-A, 100 hp., 2750 rpm. maximum. Fuel: 80/87, 26 gal. (22.5 usable). Propellers: Sensenich M69CK, McCauley 1A100/MCM. Maximum weight: 1,500 lb.

CESSNA 150D (1964), 150E (1965), 150F (1966).

Engine: Continental 0-200-A, 100 hp., 2750 rpm. maximum. Fuel: 80/87, 26 gal. (22.5 usable). Propellers: Sensenich M69CK, McCauley 1A100/MCM. Maximum weight: 1,600 lb.

CESSNA 150G (1967), 150H (1968), 150J (1969), 150K (1970).

Engine: Continental 0-200-A, 100 hp., 2750 rpm. maximum. Fuel: 80/87, 26 gal. (22.5 usable landplane, 21.5 usable, seaplane). Propellers: Sensenich M69CK, McCauley 1A100/MCM, McCauley 1A90/CF (seaplane), McCauley 1A101, DCM. Maximum weight: 1,600 lb. (seaplane 1,650 lb.) Airspeed limits: never exceed, 141 knots calibrated.

CESSNA A150K Aerobatic category. (1970).

Engine: same as 150K. Fuel: 80/87, 26 gal. (22.5 usable). Propeller: 1A101/DCM. Maximum weight: 1,600 lb. Airspeed limits: never exceed, calibrated, 168 knots.

CESSNA 150L (1971)

Engine: Continental 0-200-A, 100 hp., 2750 rpm. maximum. Fuel: 80/87, 26 gal. (22.5 usable). Propellers: McCauley 1A101/GCM, McCauley 1A101/HCM, McCauley 1A101/PCM, McCauley 1A102/OCM. Maximum weight, 1,600 lb.

CESSNA 150M (1975)

Engine: Continental 0-200-A, 100 hp., 2750 rpm. maximum. Fuel: 80/87, 26 gal. (22.5 usable). Propellers: McCauley 1A102/OCM. Maximum weight: 1,600 lb. Airspeed limits: never exceed, indicated, 141 knots.

CESSNA A150M, AEROBAT (1975).

Engine: same as 150M. Fuel: same as 150M. Propeller: same as 150M. Maximum weight: same as 150M. Airspeed limits: never exceed, 164 knots indicated.

CESSNA 152 (1978).

Engine: serial number 15279406 through 15285594. **Lycoming 0-235-L2C**; serial number 15285595 and on, and aircraft reworked per SK152-15 or SK152-16, Lycoming 0-235-N2C, 2550 rpm. 110 hp. for the 0-235-L2C and 108 hp. for the 0-235-N2C model. Fuel: 100LL/100, 26 gal. (24.5 gal. usable). Propeller: McCauley 1A103/TCM6958. Maximum weight, 1,670 lb. (1,675 lb. ramp weight for serial number 15282032 and on). Airpseed limits: never exceed, 149 knots indicated.

CESSNA A152 AEROBAT (1978).

Engine: same as 152. Fuel: same as 152. Propeller: McCauley 1A103/TCM6958. Maximum weight: same as 152. Airspeed limits: never exceed, 172 knots indicated.

150/152 Airworthiness Directives

This is only a partial list. However, it gives the potential buyer an idea of what kinds of trouble the aircraft has had. Assurance that the aircraft you purchase complies with all ADs is best given by a mechanic after thorough research into the aircraft logbooks.

Some of these apply only to certain serial numbers of a particular model (150 or 152). Again, let your mechanic check to see if the AD applies.

What has been done in this section is: (1) travel to the FAA Flight Standards District Office, and (2) sit for hours in front of their microfiche machine, printing copies of the records. Here are the 150/152 ADs selected for you.

73-23-07.

Cessna 150. To prevent defective spar attachment fitting from remaining in service, replace wing attach fittings.

76-06-02.

Cessna 150s using Avcon Industries kits incorporating defective mufflers. To prevent possible leakage of carbon monoxide into cabin heater system.

75-15-08.

Several models of 150s modified with STC (Supplemental Type Certificate) SA2219WE. To prevent loss of engine oil, improper engine lubrication or engine oil contamination.

76-01-01.

Cessna 150K and 150L, modified by STC SA1809WE, incorporating Flint Aero auxiliary tanks in the outboard wing panels. To prevent the possibility of encountering high speed flutter with fuel in these tanks.

77-02-09.

Cessna 150 (certain serial numbers). If date code stamp on the flap actuator is OH, HH, WH or ZH, flaps must not be used until the wing flap actuator ball nut assembly has been replaced.

78-25-07.

Cessna A150M and A152 (certain serial numbers for both). To assure necessary structural integrity of the vertical fin attachment to the airplane, replace vertical fin attach brackets.

79-02-06.

Cessna 152 and A152 (Aerobat) having Cessna original exhaust systems. To preclude contamination of cabin heater air with carbon monoxide.

79-08-03.

Cessna 150, A150. To prevent in-flight electrical system failure, smoke in the cockpit and/or fire in the wire bundle behind the instrument panel.

79-10-14.

Cessna 150, A (Aerobat)150. To provide an alternative source of fuel tank venting in case of fuel tank vent obstruction by foreign material and/or sticking of the fuel tank vent valve.

80-06-03.

Cessna 150M, A150M, 152, A152. To assure continued structural integrity of the wing flap direct cable, thereby preventing possible sudden unexpected retraction of the left wing flap.

80-11-04.

Cessna 150, 152, A150, A152, several model years. To detect cracked NAS 1068A4 nutplates which, if allowed to go undetected, **could result in separation of the vertical or vertical and horizontal tail assembly from the airplane.** Must be done every 100 hours for the life of the airplane. Using a light and mirror, inspect the eight nutplates.

81-05-01.

Cessna 152, A152. To reduce the possibility of fuel depletion due to incorrect fuel quantity markings.

83-17-06.

Cessna 150D, L, modified per Robertson STC SA 2191WE and SA 2192WE. To prevent possible destructive aileron flutter.

83-22-06.

Cessna 152, A152. To prevent possible loss of an aileron hinge pin.

86-15-07.

Cessna 150, 150A, B, C. To assure operation

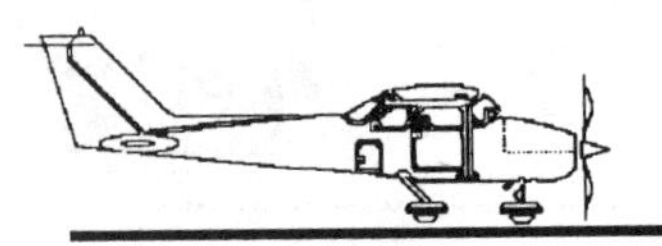
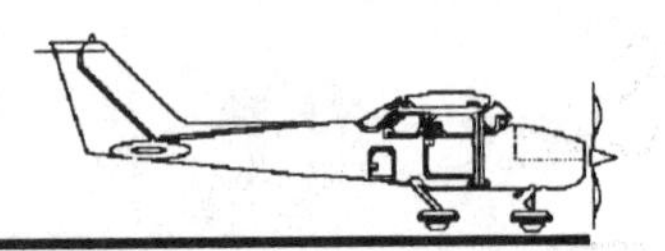

of the airplane within the approved center of gravity, weigh the airplane with full oil and unusable fuel, then add pilot weight and calculate.

86-24-07.
Cessna 150C, A150K, 152, A152. To prevent engine power interruption due to loss of attachment of the engine controls.

86-26-04.
Cessna 150, many models. To prevent slippage of the pilot/co-pilot shoulder harness.

No Nasty 150 Service Bulletins

There are no huge problems listed among the Cessna 150 service bulletins. For the 150, the most recent bulletin was 1986 and concerned the shoulder harness adjuster spring inspection and removal. Prior to that, going back to 1978, service letters concern: (SE-78-1) nut plate inspection of the vertical fin attachment bracket, a second letter about the same bracket, (SE 78-72) tail cone corrosion, a third letter about nut plate inspection of the vertical fin attachment bracket, (SE 80-58) improved window and windshield sealer, (SE 83-6) seat rail inspections which affected all Cessnas (1983), and (SNL 85-18) shoulder harness kit availability.

The most recent letter issued against the landing gear was 1984 (SE 84-21) and concerned nose wheel shimmy troubleshooting.

The last four service letters about the fuel system start at 1982 and go back to 1979, and concern: (SE 82-36) fuel contamination, and the installation of quick drains to fix the problem (SE 79-45).

Here are engine service letter topics from 1986 back to 1978:(SNL 86-39) Phillips XC 2 Multi-Viscosity oil, (SNL 85-34) proper engine break-in procedure for Continental engines, (SNL 85-8) oil leak troubleshooting, (VSBN 79-9) Continental M79-7 crankcase identification, (VSBN 78-6) Continental M78-4 — availability of silk thread, (VSBN 78-1) Continental M76-19 R1 — EGT recommendations, and (SE 78-67 R1) oil quick drain.

Electrical system service letters start in 1980. This review takes us back to 1977: (SE 80-86) Electrical System Improvements Since 1976, (SE 80-52) flashing beacon power supply, (SE 78-18) external power receptacle wiring, (SE 77-21) aircraft battery electrolyte.

Ditto For 152 Service Bulletins

There are few service bulletins between 1983 and 1986 for the 152, and those are minor. Here are bulletins for the fuselage and airframe: (SNL 86-44) flight control surface balancing information, (SE 84-22) aileron hinge assembly improvement, (SE 83-81) rudder bar weld assembly improvement, (SE 83-26) horizontal stabilizer and elevator improvement, and (SE 83-18) aileron hinge pin inspection.

It is more important to look at the engine bulletins for the 152, since it has a Lycoming and the 150 has a Continental engine. There again, we find few problems. In fact, from 1986 back to 1983 there were few directly related to the 0-235 engine. Here's a sample: (SNL 86-39) Phillips XC 2 Multi Viscosity oil, (SNL 86-31) increased TBO for Lycoming 0-235 engines (hardly a problem), (SNL 85-34) proper engine break-in procedure, (SNL 85- 8) oil leak troubleshooting maintenance tip, (SE 84-2) maintenance procedures to reduce valve sticking in Lycoming engines.

To find anything halfway serious you have to go back to 1980, when there was a bulletin on

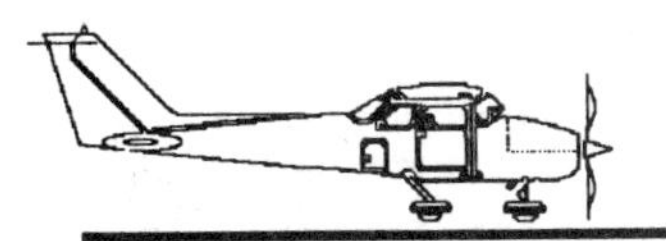

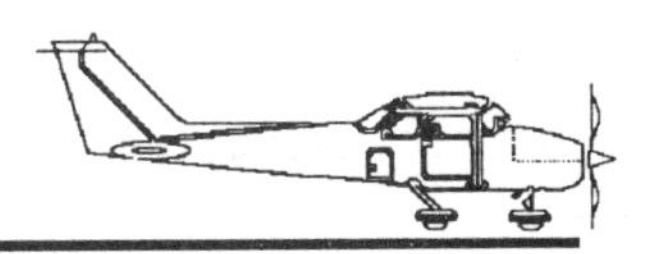

push rod inspections, and again in 1980 when there was a bulletin about inspection of crankshaft flanges.

152 Accident Reports

A quick look at National Transportation Safety Board (NTSB) accident reports for 1986 shows the Cessna 150/152 runs out of fuel a heck of a lot more than it breaks.

There was the CFI who took off with 12 gallons, the private pilot who took off with six and the student who took off with a few pints during a cross country, despite advice from his instructor to refuel at intermediate stops.

Even when they do fail, poor maintenance (loose fuel lines) or no maintenance often was the culprit. However, we include those in this analysis of 18 mechanical failures — all the ones there were for 1986. (In one case a mechanic jammed an oversize part into the engine.)

As with other NTSB records, there were obvious pilot-error accidents that also had "mechanical" as a cause. Nearly every time a pilot ran out of fuel, he broke something (otherwise the NTSB would not be notified) and that triggered the assigning of an additional cause in NTSB records — a listing of "mechanical." Those have been thrown out of this list.

Jan. 2, 1986
The number one intake valve of a Cessna 150L stuck in the open position, causing a partial loss of power leading to a precautionary landing.

Jan. 13
The instructor stated flaps on a 152 would not retract on a touch and go and the aircraft landed off airport near Essex County Airport, N.J.

Jan. 14
The exhaust valve rocker arm for number one cylinder on a 150G failed in flight and later examination showed signs of fatigue prior to failure. Student landed in woods when he could have landed on a frozen lake.

April 3
A nut securing the nosewheel axle of a Cessna 150H "backed off," allowing the wheel to become misaligned. The aircraft veered into a ditch.

April 4
An airworthiness directive to change the air cleaner element on a Cessna 150J every 12 months or 100 hours was ignored, and the foam air cleaner was drawn into the carburetor air box, blocking the air intake to the carburetor.

April 22
Erroneous airspeed indications caused a Cessna 150D to float excessively on landing, leading to a hard bounce when the pilot raised the flaps in an effort to get the plane on the runway.

May 11
Student pilot attempted to start a Cessna 152 in which the carburetor had been removed for maintenance, despite written and verbal warnings. The engine caught fire.

May 30
Private pilot failed to remove air intake plugs used to keep birds out, and engine overheated during flight, leading to a forced landing.

May 31
Right tire of a Cessna 150 apparently deflated in flight, causing the aircraft to swerve and hit a parked Cessna 172 on landing.

June 2
Commercial pilot washed the aircraft, getting

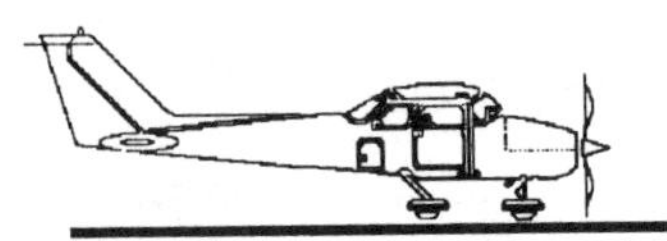
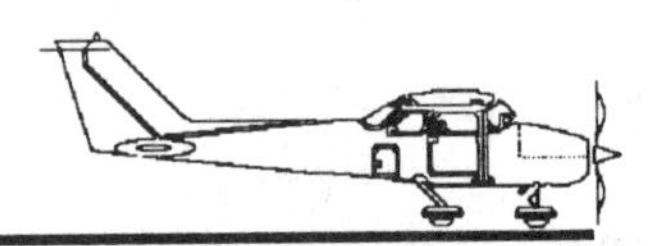

water in the magnetos, then flew it. The magnetos cut out, leading to engine failure.

June 21

The number two cylinder connecting rod failed and penetrated the engine case of a Cessna 150M. The crankshaft journal for the number two cylinder was heat damaged and the rod cap bearing could not be found. Other cylinder bearings were scored and extensively worn. The engine had been overhauled 832 hours previously. Two minor injuries at Brandon, Fla.

July 23

A magneto ignition system malfunctioned due to excessive wear in a Cessna 150 during the first banner two of the day, causing the aircraft to descend into the trees after dropping the banner.

July 31

A private pilot took off in a Cessna 150H on a test fight after an annual inspection and top overhaul at Bartlesville, Okla. The woman landed in dense brush and trees with minor injuries. The number three exhaust valve guide was an incorrect part, and was oversize in length by .165 in. It would not allow full valve movement, and the residual movement resulted in failure of the rocker arm support structure.

Aug. 4

A commercial pilot of a Cessna 150L made a precautionary landing after loss of rpm. Examination of the engine revealed a coating of carbon and lead on the valves of number two, three and four cylinders which caused the valves to stick in the open position and bind.

Sept. 2

During an instructional flight a partial power loss occurred due to the mixture control cable at the carburetor end becoming loose. During the forced landing the Kent, Wash., the pilot struck two railroad ties on a dirt pile.

Sept. 10

The commercial pilot of a 152 got carburetor ice, applied heat and regained power, then made a precautionary landing and checked the engine. He took off again, climbed to 3,500 ft. and got carb ice again, but the carb heat control was stuck. He lost power and landed.

Oct. 8

The engine of a Cessna 150L quit on final due to a broken mixture control cable housing. The private pilot had just reduced power and lowered 10 degrees of flaps for landing.

Nov. 9

The blade of a Cessna 152 propeller at Conway, S.C., separated three inches from the hub during an instructional flight due to fatigue cracks originating from chordwise surface scratches underneath the painted finish. The propeller time in service, without refinishing, was 3,425 hours.

While there were other 150/152 accidents in 1986, records indicating causes show entries such as "preflight, inadequate" and "fuel, exhaustion." In one case, a larger 0-320 series engine meant for a 172 had been placed on a 152, and the student used book values for the fuel consumption of a standard 0-235 engine. He ran out of fuel. While NTSB officially blames the plane in those cases, we don't.

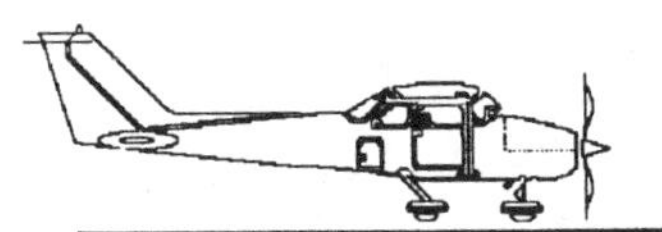
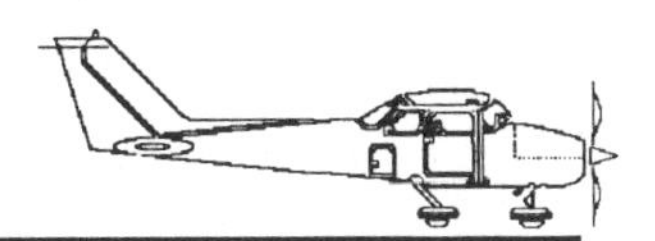

CHAPTER FIVE

THE CESSNA 182

How Much 182 Can You Afford?

The main thing you would like to know is the skill of the pilot you bought it from. Skylanes, with one pilot aboard, are a little nose heavy. That can easily be compensated for by skill and use of full nose up trim at touchdown. It is not a difficult plane at all.

If the previous owner was unskilled, maybe even a klutz, the aircraft may have a wrinkled firewall and there may be repair jobs in the logbook for prop strikes due to nosewheel landings. If there is nothing in the logbook and the firewall looks OK, then the previous owner probably exercised normal caution.

The other, obvious thing to watch with a 182 is water in the bladder fuel tanks. Planes have arrived at a dealer for trade-in with gallons of water caught in the wrinkles of the rubber fuel cell. **Shake, shake, shake your bladder.** There was once an airworthiness directive asking owners to "rock wingtips vigorously."

An examination of a recent issue of *Trade-A-Plane* advertised 187 Skylanes, including retractable gear, normally aspirated and turbo models, but only seven of those were turbo. The turbo is a very popular model, and holds its value well. That is why a 1968 Skylane turbo still has an asking price of $27,500, and a 1981 model is priced at $66,000. Drop back three years, to 1978, and the turbo price is still $58,000.

The Skylane is a favorite of mechanics. They said, during interviews, they feel it is one of the best-made Cessnas ever produced. (See "The Infamous 'H'" section.) It has an excellent speed advantage over the Cessna 172. Marketing hype aside, you will actually get about 100-105 knots in a Cessna 172 and 135 in a 182.

If you want one cheap, try the 1975 model with an average asking price, according to the *Trade-A-Plane* ads, of $33,000. After that the price climbs rapidly, to over $50,000 for a 1980 model.

You have a choice in prices, since there are

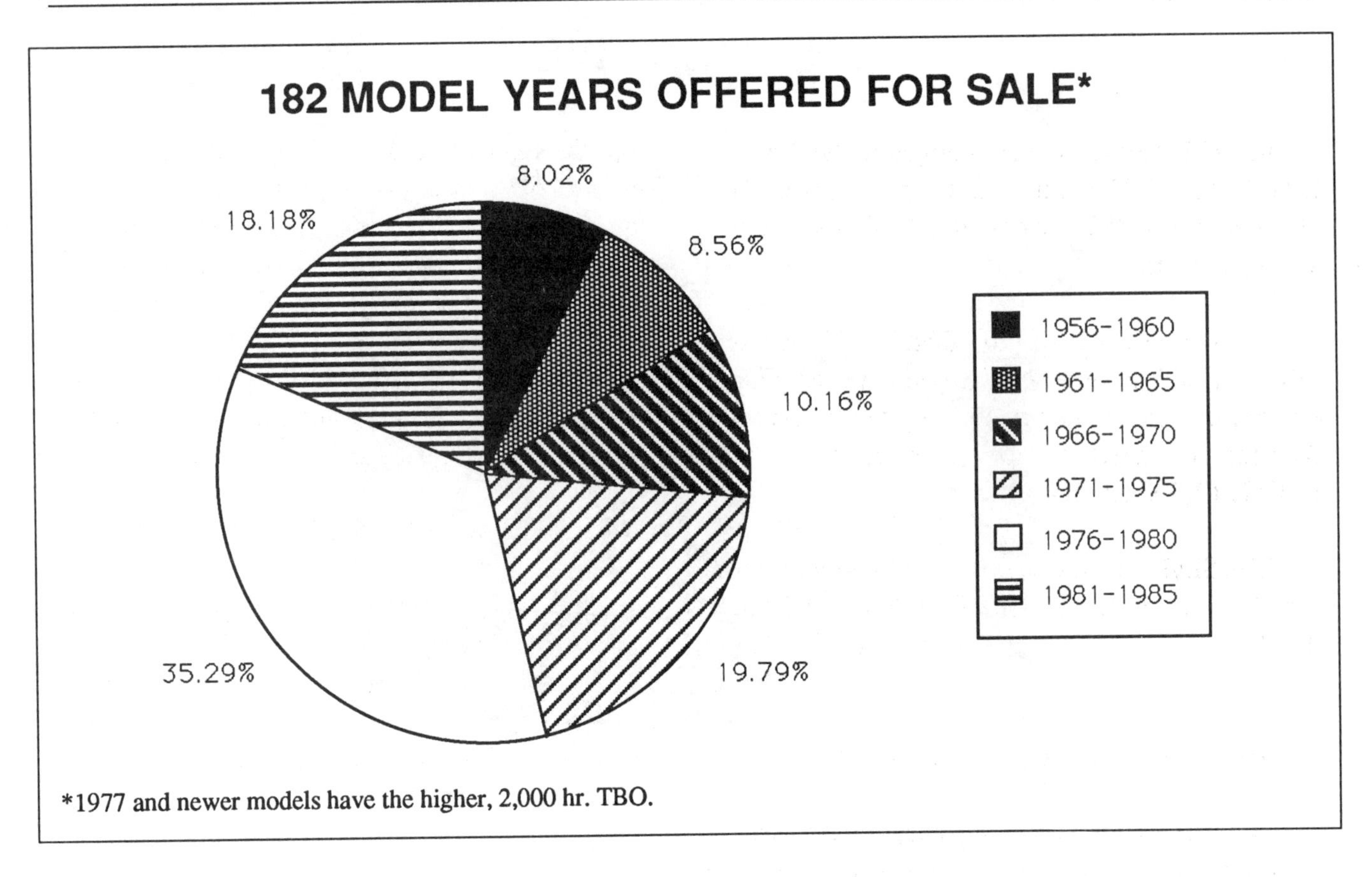

fixed-gear and retractable gear Skylanes. Obviously, operating costs are lower for the fixed-gear aircraft.

A few owners have complained about engine noise in the Skylane, even with a pair of David Clark headphones clamped over their ears.

The 182 began in 1956 with the Continental 0-470-L engine (1,500 hr. TBO) at 230 hp.

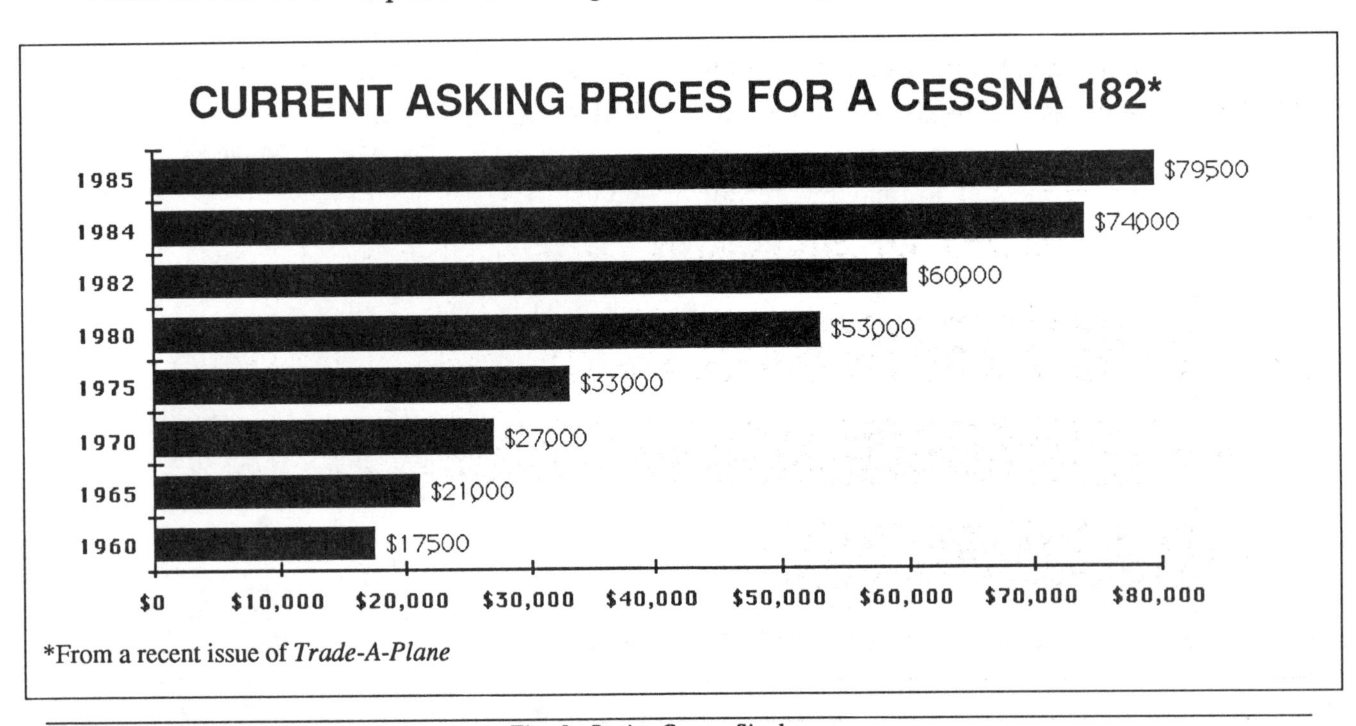

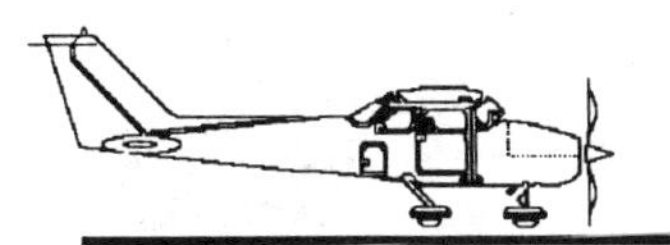
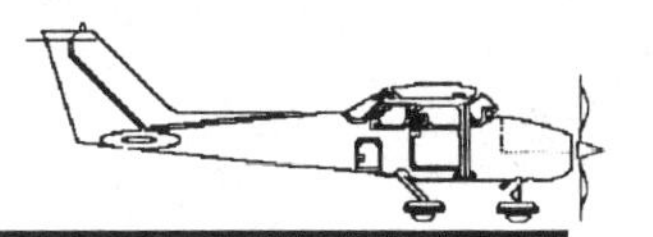

In 1962 the engine was changed to the 0-470-R (1,500 hr. TBO), still at 230 hp., and was not changed until 1975 when the 0-470-S (same TBO) was used.

In 1977 the 0-470-U (2,000 hr. TBO if new or rebuilt, 1,500 hr. otherwise) became the Skylane engine, and was joined in 1981 by the first turbo Skylane powered by the Lycoming 0-540-L3C5D (2,000 hr. TBO).

The Skylane RGs began in 1978 powered by the Lycoming 0-540-J3C5D (also 2,000 hr. TBO).

Annual Inspections: A 182 And An RG

Here are maintenance costs paid by the owner of a Skylane compared to those of a Skylane RG over a recent six-month period. As you will see, the fixed-gear annual inspection actually costs more than that for the Skylane RG. Both aircraft are operated privately. These are actual aircraft, now flying on East Coast.

One important trend should be noted. Both owners had gone to a variety of maintenance shops for repair work, including some bad ones. The bad ones made poor adjustments to parts requiring exact tolerance, and even installed some parts wrong. The good shop then had to repair the improper work and charge for it. In most cases, the customer wouldn't believe the parts were installed wrong. Go to the right shop in the first place, and keep going there even though it costs more. It costs less in the long run.

First A Look At The Straight Leg 182

Early in the six-month period the nose strut went flat, and the cost of repair was $71.37. A few months later a cowl flap was found missing, and it took $70 in labor to put on a new $470 cowl flap. The total bill that time was over $550. Then it was time for the annual.

The annual inspection cost $3,500, which included $2,500 in labor. When the owner brought the airplane to the shop, he stated there was nothing wrong, except for the nose strut, which was still acting up, and an airworthiness directive which had come out a few months earlier. The airworthiness directive concerned the throttle linkage and cost only $30.

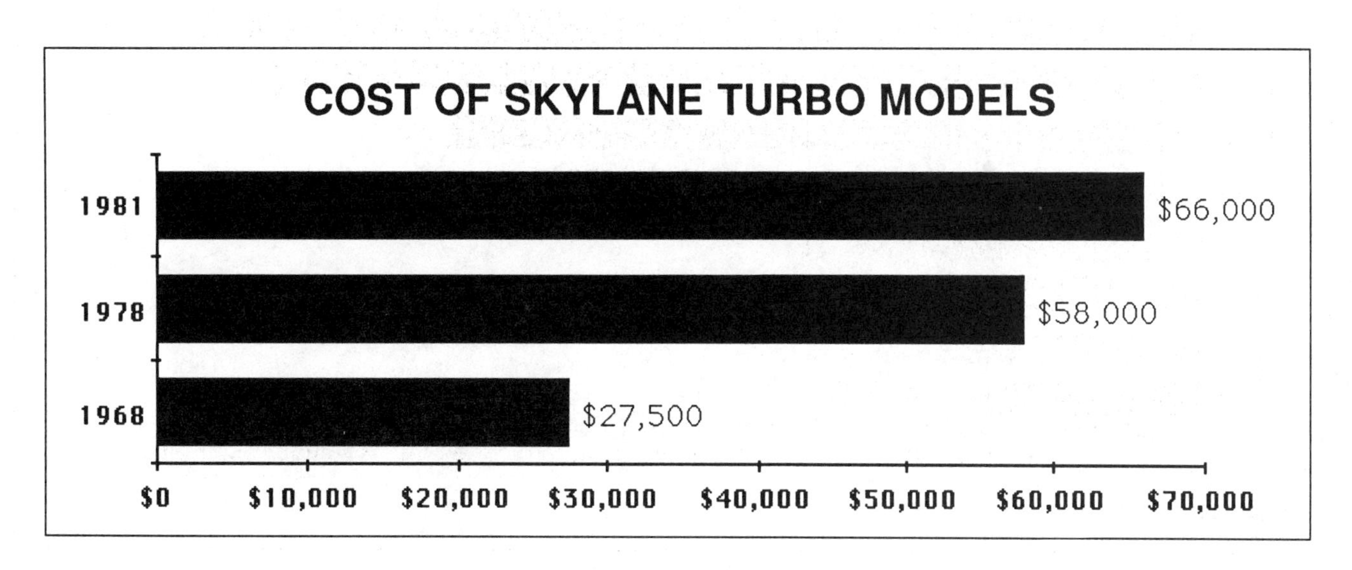

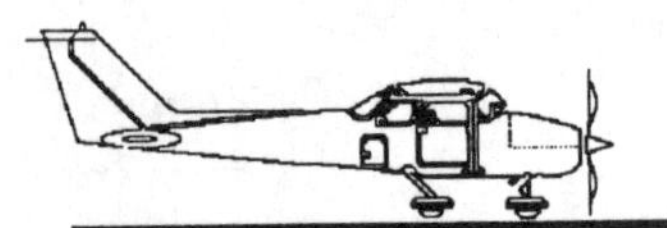

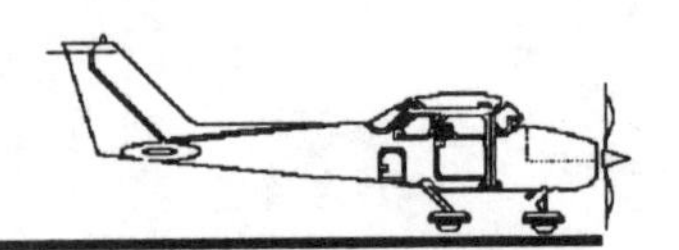

Included in the annual were these costs.

> A cylinder showed low compression, and cost $500 to repair.

> The nose strut cost another $44.

> Two engine baffles were cracked, costing $15 and $60 to repair.

> The hose to a brake cylinder was leaking, costing $60.

> Aileron cable tension was low, costing $60.

> A set of rings cost $70, while an ELT battery was $30, points were $35 and 12 quarts of oil were $30.

> Wing cracks cost $85 and $100 to repair.

> The seat rail was cracked and cost $80.

> The carb heat cable was worn and cost $60.

> The trim light bulb and a fire extinguisher mount both cost $25 to repair.

> A landing light, one of two, cost $30, while pitting to the magneto points cost $85 to replace.

Skylane RGs Actually Cost Less

Now let's compare thoses costs with a Skylane RG. This aircraft had only 700 hours on it, and seems to prove that newer aircraft cost less to maintain. The annual was $1,600, and includes a higher basic fee than charged for the fixed gear 182.

Before the annual, the aircraft made a visit to

Sample Yearly Insurance Rates For A Cessna 182*

Year	Value	Pvt. 250 hr.	Pvt. 50 csp** 500 hr.	Pvt. 100 csp** 1,300 hr.
1960	$17,500	$1,200	$1,100	$1,000
1970	27,000	1,320	1,200	1,100
1975	33,000	1,425	1,305	1,210
1980	53,000	1,615	1,512	1,425
1985	79,500	1,900	1,750	1,600

* 15% more for a 182RG (retractable gear).
No difference in premium for turbo 182.
The liability limits are based on $500,000 combined single limit (includes bodily injury to the public, property damage liability and passenger liability limited to $50,000 each person).

** csp refers to constant speed prop experience.

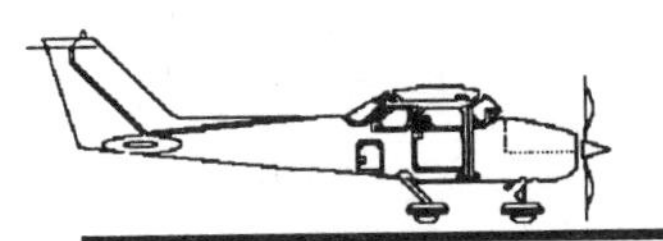
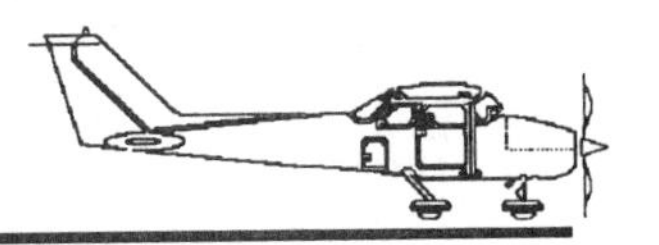

the shop to comply with an airworthiness directive on seat rails, which cost $36. A second visit repaired the starter, changed the oil and took care of other minor routine items for $600.

The annual included these problems:

> The autopilot roll servo motor required repair, at $100.

> Wheel bearings in the nose and main gear were worn, and each cost $28.

> The seal on a fuel tank cap deteriorated, and cost $45 to repair.

> A fuel tank filler neck was corroded and the cap was broken, costing $40 to repair.

> Inspection panel screws were corroded and cost $30.

> A bird built a nest in the plane, and it cost $25 to remove it.

> A cylinder intake pipe gasket was leaking, and that cost $30.

> The carburetor heat duct was **improperly installed**, and cost $10 to reinstall.

So, you don't necessarily save on an annual for a fixed-gear 182 compared to a 182RG. There is, however, more to go wrong on an RG. Catch problems early, keep the plane in good repair, don't "fool" yourself that everything is working well, and those annual inspections could be cheaper.

It is also evident that owners must do a much more thorough preflight than renter pilots. Owners might keep maintenance costs down by incorporating several steps from the **World's Best Preflight**, found in this book, every time they fly.

The 182 Type Certificate

The Cessna 182 type certificate runs 26 pages. As always, the thing you have to be concerned about are the notes at the end of the certificate.

Other than those, you must know what is supposed to be on the airplane you buy, and check to see all parts match the type certificate. If they don't, there must be a supplemental type certificate approving the parts that don't match the type certificate. No supplemental type certificate, no flying.

There are no serious notes on the 182. Note one requires that certain weight and balance numbers be used for unusable fuel and oil on certain serial numbers. Note two talks about what placards must be posted in full view of the pilot. Note three is more interesting, requiring thermisters to be installed in exact cylinder locations specified on 1970 through 1980 models. Your mechanic should be aware of this one.

Note four says using the 0-470-S engine in the Cessna 182N or P models requires a change of the oil temperature gauge. That takes in all 1970 through 1975 models. Note five concerns some more placards. Note six tells you that — through 1977 — the airplane had a 14-volt electrical system, but thereafter it had a 28-volt system.

CESSNA 182 (Model Year 1957)

Continental 0-470-L engine, 2,600 rpm. (230 hp.). 80 minimum grade gasoline, 60 gal. (55 gal. usable). Maximum weight, 2,550 lb. Propeller approved: Hartzell hub HC82XF-1 or HCA2XF-1 or BHCA2XF-1 with 8433-2 blades, McCauley hub 2A36C with 90M-8 blades, Hartzell hub

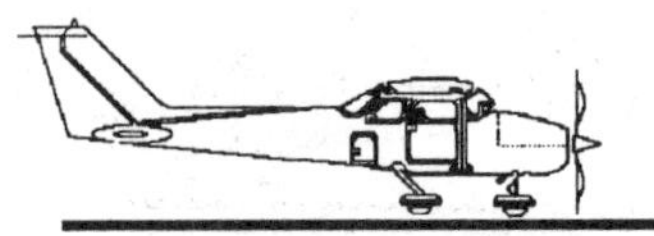
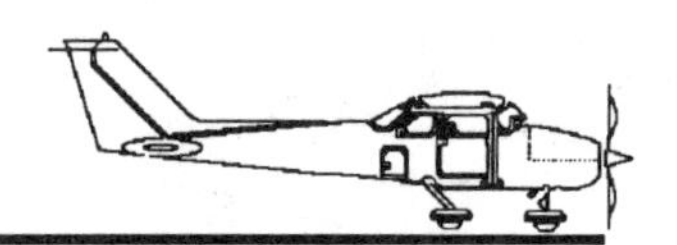

BHC-C2YF-1 with 8468-2 blades, McCauley hub 2A34C with 90A-8 or 90AT-8 blades. Max cruise, 139 knots.

CESSNA 182A (Model Year 1957).

As above, except: Maximum weight, 2,650 lb., fuel 65 gal., 55 gal. usable.

CESSNA 182B (1959).

Continental 0-470-L engine, 2,600 rpm., (230 hp.). 80 grade gas, 65 gal. (55 gal. usable). Maximum weight 2,650 lb. Propellers as in Cessna 182 above. Max cruise, 139 knots.

CESSNA 182C (1960), 182D (1961).

Continental 0-470-L, 2,600 rpm. (230 hp.). 80 grade gas, 65 gal. (55 gal. usable). Maximum weight, 2,650 lb. Propellers as with Cessna 182 above. Max cruise 139 knots.

CESSNA 182E (1962), 182F (1963), 182G (1964).

Continental 0-470-L or 0-470-R, 2,600 rpm. (230 hp.). 80/87 grade gas, 65 gal., **60 usable. Maximum weight, 2,800 lb.** Propellers same as Cessna 182. Max cruise 139 knots.

CESSNA 182H (1965), 182J (1966), 182K (1967), 182L (1968).

Continental 0-470-R, 2,600 rpm. (230 hp.). 80/87 grade gas, 65 gal. (60 usable). Maximum weight, 2,800 lb. Propeller, McCauley hub 2A34C66/90AT-8 blades. Max cruise, 139 knots.

CESSNA 182M (1969).

Continental 0-470-R, 2,600 rpm., (230 hp.). 80/87 grade gas, 65 gal. (60 gal. usable). Maximum weight, 2,800 lb. Propellers, McCauley hub 2A34C66/90AT-8 blades, McCauley hub 2A34C201/90DA-8 blades, McCauley 2A34C203/90DCA-8 blades. Max cruise, 139 knots.

CESSNA 182N (1970).

Continental 0-470-R or 0-470-S, 2,600 rpm. (230 hp.). 80/87 grade gas, 65 gal. (60 gal. usable). **Maximum weight, 2,950 lb.** takeoff, 2,800 landing. Propellers, McCauley 2A34C201/90DA-8 blades, McCauley 2434C66/90AT-8 blades, McCauley 2A34C203/90DCA-8 blades. Max cruise 139 knots.

CESSNA 182P (1972).

Continental 0-470-R (serial number 18260826 through 18263475), 0-470-S (serial number 18260826 and up), 2,600 rpm. (230 hp.) 80/87 gas, 65 gal. (60 usable) or long range tanks 80 gal. (75 gal. usable). Maximum weight, 2,950 lb. Propellers as with 182N. Max cruise, 139 knots.

CESSNA 182Q (1977).

Continental 0-470-U, 2,400 rpm. (230 hp.). 100/130 grade gas (serial number 18265176 through 18265965), **100LL/100 grade gas (serial number 18265966 through 18267715), 61 gal. (56 gal. usable)**, long range 80 gal. (75 usable) and 92 gal. (88 usable). Maximum weight, 2,950 lb. Propeller, McCauley hub C2A34C204/90DCB-8 blades. (No other props approved.) **Max cruise, 143 knots.**

CESSNA R182, SKYLANE RG (1978), TR182 TURBO SKYLANE RG (1979).

Model R182, Lycoming 0-540-J3C5D, 2,400 rpm., (235 hp.). Model TR182, Lycoming 0-540-L3C5D, 2,400 rpm. 31 in. hg. (235 hp.) 100LL/100 grade gas, 61 gal. (56 usable), long range 80 gal. (75 usable), 92 gal. (88 usable). Maximum weight, 3,100 lb. Propellers, McCauley hub B2D34C214/90DHB-8 blades, McCauley hub B2D34C218/90DHB-8 blades, McCauley hub B3D32C407/82NDA-3 blades. Max cruise, 1978 R182 — 143 knots, 1979 R182 — 160 knots, TR182 — 157 knots. 1980 model and up, R182—

159 knots, TR182 — 157 knots.

CESSNA 182R (1981), T182 (1981).

Model 182R, Continental 0-470-U, 2,400 rpm. (230 hp.), Model T182, Lycoming 0-540-L3C5D 2,400 rpm., 31 in. hg. (235 hp.). 100LL/100 grade gas, 92 gal. (88 gal. usable). Maximum weight, 3,100 takeoff and 2,950 landing. Propellers, McCauley hub B2D34C219/90DHB-8 blades, McCauley hub B3D32C407/82NDA-3 blades. Max cruise, Model 182R — 143 knots, T182 — 140 knots.

A Talk With A Dealer

Mid West Aircraft, Sandwich, Ill., has discovered a powerful new marketing tool in the sale of used aircraft: quality.

They discovered people were hungry for quality used aircraft they could trust. So Mid West buys first-time run-out aircraft — a lot of them Cessna 182s — and overhauls the engine and prop. Then they paint it if needed and renew the upholstery as necessary. They don't do "silly" things, like replace all the cable pulleys for no reason. The price isn't low, so they put prices in their ads to "pre-qualify" customers.

"We don't broker any airplanes," Mid West sales manager Donn Lynch said. They feel they need to own the airplane in order to have the authority to repair it as needed.

Consequently, they have a lot of experience with 182s, not only in sales but in their mechanical performance, since they pull the engine on every one that comes in.

Be Aware Of The Limitations

Lynch was asked for problems with the 182, and gave a candid answer. But he also pointed out that anyone who is aware of the limitations and accepts the responsibility for knowing how to take care of the airplane can easily avoid trouble.

First, everyone knows about undrainable water in the bladder fuel tanks. Vigorous rocking of the wings, if you suspect water, often helps.

Lynch said the Continental engine is, under the right conditions, prone to carburetor icing. Skylane pilots are familiar with flying along smoothly, then applying carb heat for a descent or a maneuver and getting momentary rough engine operation. Lynch said that is due to the carb ice melting and sending a slug of water through the engine. He has flown them to Europe and Australia with no problems, but due to his awareness of potential icing he pulled the carb heat every 10 minutes during one ferry flight after encountering a rainstorm, he said.

The Secret Of The Skylane

Another tip comes in the area of weight and balance. With only one person aboard, the aircraft tends to be nose heavy. Consequently, a lot of pilots who are unaware of the characteristic land on the nose wheel. Therefore it is important when checking 182 logbooks to look for prop strikes and firewall damage. When inspecting the aircraft, be sure to check visually for wrinkled firewalls. The answer, Lynch said, is to do what most pilots of heavier aircraft such as the Cessna 210 and twins do; roll full nose up trim just prior to touchdown.

If a repair is listed in the logbook as "details on file," get those details from the repair station. Lynch recalled a twin Comanche which had low time but listed repair details as being "on file."

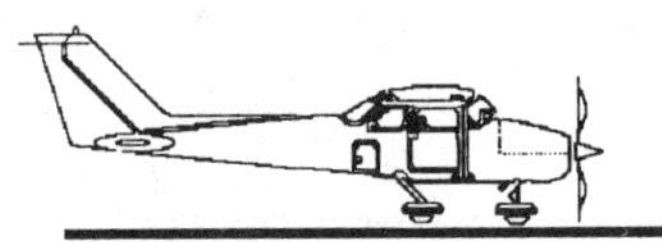
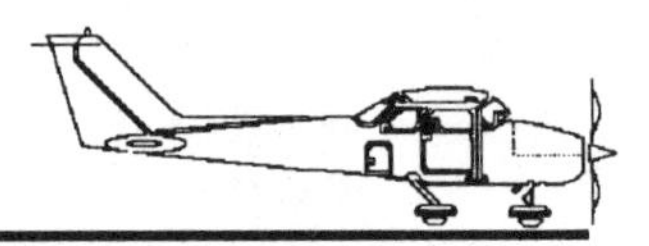

When the shop was called, it was learned the airplane had crashed, and its wings sawed off to get it on a truck, before being donated to a mechanic training school where it sat for years while students repaired it. Interesting history.

He said prop overhauls, normally needed at 1,500 hrs. or five years, can cost as little as $1,000.

Lynch, a retired Chicago air traffic controller, said Cessna Finance is having to set up a separate policy for Mid West Aircraft. Cessna Finance will only loan 80% of Blue Book price, and Mid West prices are routinely higher because their planes have so much refurbishment done prior to sale. In the interim, Cessna Finance has decided to finance 85% of Blue Book price.

Mid West picked the 182 because they wanted a popular airplane. They avoided Beech aircraft because prices, after all the refurbishment, would be too high.

The bottom line: Mid West sells lots of 182s because of the plane's reputation for lifting whatever asked, safely. But accept the responsibility for knowing how to take care of the airplane, and you won't be disappointed.

182 Airworthiness Directives

This is not the complete airworthiness directive list for the Skylane, but it gives you a good idea of what sort of trouble the aircraft has had. Your mechanic will research the entire list and see whether the aircraft you pick has all the directives performed. This partial list is what the FAA calls "Book Two," which contains ADs from 1971 on.

Many of the ADs apply to more than one Cessna model, and in some cases more than one brand of aircraft. Here is a look at the last 17 years of 182 ADs.

75-05-02.

Affects 15 models of Cessna 182s, covering any 182 modified in accordance with STC (supplemental type certificate) SA2653WE. To prevent loss of engine oil, improper engine lubrication or engine oil contamination.

75-16-01.

Cessna 182. To preclude inadvertent fuel exhaustion due to incorrect fuel placard capacities.

76-04-03.

Cessna 182N, P, having ARC PA-500A actuators installed as part of Cessna model 300, 400 and 400A autopilots, 300, 400 or 800 IFCS, or in type G-830A yaw damper systems. To preclude restrictions of control movement due to jamming of the ARC PA-500A actuator gear train.

77-02-09.

Cessna 182 (certain serial numbers). If date code stamp on the flap actuator is OH, HH, WH or ZH, flaps must not be used until the wing flap actuator ball nut assembly has been replaced.

77-12-08.

Cessna 182P, Q, equipped with electrical ground power receptacles. To prevent unwanted propeller rotation.

77-14-09.

Cessna 182P, Q. To assure aircraft controllability when operating at maximum gross weight and aft center of gravity location, limiting aft CG location to 46 inches.

77-23-11.

Cessna 182 with certain ELT installations. To preclude the possibility of an in-flight fire due to a loose ELT antenna coaxial cable connector making contact with the terminals of the battery

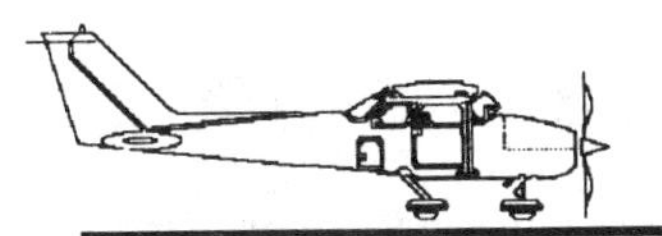
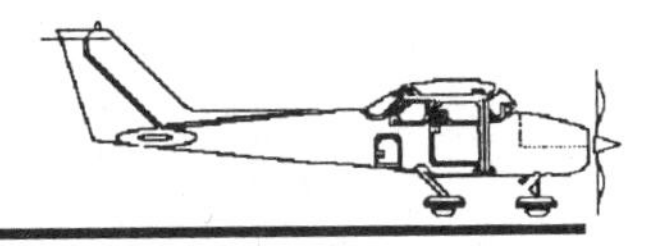

relay and thereby shorting the electrical power to ground.

79-08-03.

Cessna 182. To prevent in-flight electrical system failure, smoke in the cockpit and/or fire in the wire bundle behind the instrument panel.

790-10-14.

Cessna 182. To provide an alternative source of fuel tank venting in case of fuel tank vent obstruction by foreign material and/or sticking of the fuel tank vent valve.

79-25-07.

To preclude the possibility of electrical or electronic component damage or an in-flight fire due to a short between an ungrounded alternator and flammable fluid-carrying lines.

83-13-01.

Many models of the Cessna 182. To alert the pilot to the potential effects of improper fuel cap sealing which can cause loss of fuel and erroneously high fuel quantity indications.

83-17-06.

Cessna 182, R182, T182 and TR182 modified per Robertson STC SA1382WE. To prevent possible destructive aileron flutter.

83-22-06.

Cessna 182Q, R, T182, R182 and TR182. To prevent possible loss of an aileron hinge pin.

84-10-01.

Cessna R182, 182, A182. To prevent power loss or engine stoppage due to water contamination of the fuel system, install quick drains and check for fuel leakage.

86-24-07.

Cessna 182F through Q, R182. To prevent engine power interruption due to loss of attachment of the engine controls.

86-26-04.

Cessna 182, many models. To prevent slippage of the pilot/co-pilot shoulder harness.

Service Bulletins For The Skylane

The most recent service bulletins affected all Cessnas, not just the 182s. To find one of any significance on the airframe, we must go back to 1982. Here is a sampling of airframe bulletins: (SE 82-43) pilot and copilot seat belt modifications, (SE 80-84) improved baggage door lock cam, (SE 80-73) improved nose cap stiffeners, (SE 78-22) corrosion in the tail cone due to lithium batteries, (SE 77-11) horizontal stabilizer rear spar flange inspection.

Power plant service bulletins look like this for the last few years: (SNL 85-34) proper engine break-in procedures, (SEB 85-4) exhaust valve replacement, (SE 84-2) maintenance procedures to reduce valve sticking in Lycoming engines, (SE 82-6) crankshaft oil seal retainer.

There were few service bulletins on the prop after 1982.

The fuel system is another story. The 182 and 210 are famous for storing water in the folds of their rubber bladder fuel tanks. The service bulletins reflect this: (SNL 85-30) fuel tank access panel sealant, (SEB 85-2) fuel quantity transmitter gasket sealing improvement, (SE 84-16) R1 fuel filler diameter modification and external turbo placard removal, (SE 84-4) rubber fuel cell inspection, (SE 82-36) fuel contamination, (SE 82-34) flush fuel cap sealing test, (SE 82-33) **warning against use of automotive gasoline**. Incidentally, a good flush fuel cap sealing test is

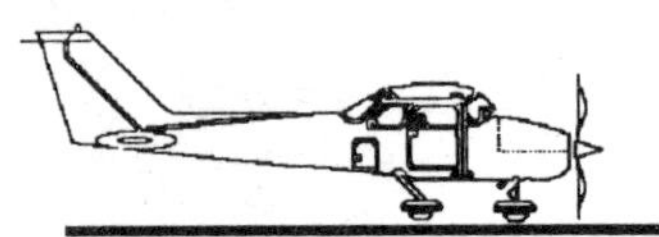 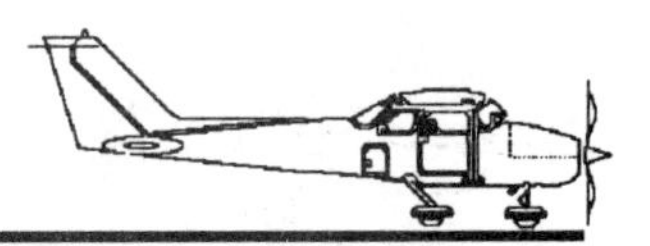

to take a cup of fuel, close and lock the fuel cap, and pour the fuel on the cap. If it enters the tank, you've got a sealing problem.

Mid West Aircraft officials said another service letter calls for opening the back of the wing and pulling out the fuel bladder wrinkles. That eliminates the need to rock the wings during preflight to discover water in the fuel, and takes care of a few worries along the way.

182 Accidents For 1986

A bogus part is included in the National Transportation Safety Board (NTSB) accident reports on Cessna 182 aircraft. Actually, it was a reworking of an original part, but an "altered" part is as good as bogus. As with other models in this book, the NTSB includes a lot of non-mechanical causes in the reports pulled from the NTSB computer, even though a request was made for only mechanical failures.

Straining those out, there were 12 purely mechanical accidents in 1986. Here's the list:

Jan. 20, 1986

A Cessna 182Q lost power on final approach after a throttle movement to correct the glidepath. The carburetor airbox which regulates the carburetor heat was improperly repaired by redrilling a hole in the control arm attachment point. The NTSB listed the part, therefore, as "bogus."

Jan. 31

Oil was lost from a 182 at Sandy, Ore., because the oil filter adapter had slipped past the threads in the oil pump housing. The aircraft had the oil filter adapter replaced as well as the O ring on Dec. 11, 1985. The mechanic had told the 39-year-old owner via a written work order that the threads on the adapter were worn and broken.

May 4

A 182D flown by a 56-year-old private pilot made a forced landing after three pieces of felt-like sealing material lodged in the carburetor venturi. It was identical to the missing seal from the junction of a ram air duct assembly and the carburetor airbox assembly. The airbox seal is the subject of an airworthiness directive (77-04-05). The AD was complied with on May 13, and no follow-up inspection was required.

May 17

Just before the descent of a 182E, the 52-year-old private pilot pushed the prop control forward and experienced a runaway prop and lost oil pressure. Then a connecting rod failed and an emergency landing was made near Crystal River, Fla. Later, no evidence of prop or oil pump failure was found.

July 28

The right brake of a 182 failed during landing roll at 25 mph, causing the aircraft to veer into a ditch. The right brake was found to be completely worn while the left had 15%-20% wear.

Aug. 19

The number one cylinder exhaust valve rocker arm shaft boss failed on a 182N. The cylinder had been changed 27 hours prior to the accident, and the rocker arm shaft keeper bolt had been over torqued during the overhaul. The aircraft struck power lines, two small buildings and a brick wall during the night forced landing, with two occupants injured seriously.

Aug. 26

The gear-down hose assembly ruptured, allowing loss of hydraulic fluid in a TR182RG at San Jose, Calif., resulting in a gear up landing on

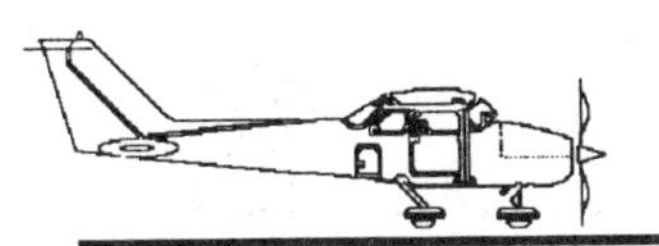
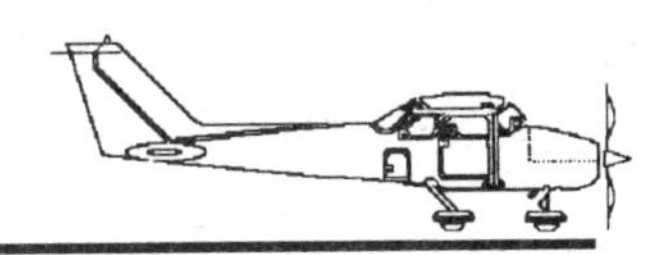

the runway.

Aug. 30

The left main landing gear spring strut, part number 0741601-1, failed where it was clamped to the fuselage of a Cessna 182F, resulting in collapse of the left main gear at touchdown. Fatigue cracking had occurred in the lower forward area of the spring strut where fretting had occurred at the clamp. The aircraft had a total flight time of 3,060 hours.

Oct. 12

A Cessna 182A cruising near King City, Calif., made a forced landing and hit a tree after the mixture control cable clamp slipped, causing the mixture to move to idle cutoff.

Oct. 22

A seat failed on a Cessna 182M at Gila Bend, Ariz., causing the pilot to slide aft where he could not reach the throttle or rudder pedals. The hardware used to assemble the fore-aft seat adjustment was incorrect. One cotter pin was installed instead of a roll pin on the lower portion of the same locking pin. The installation of the cotter pins allowed too much play, causing the fore-aft seat adjustment locking pin to fail to engage the seat rail properly.

Nov. 8

The engine of a 182B backfired during startup and an induction fire started. The pilot did not continue cranking the engine. High winds inhibited the effectiveness of the fire extinguisher. An extinguisher from the hanger did not work and the aircraft was substantially damaged.

Nov. 9

After what the pilot said was a smooth landing of a Cessna 182G at Juneau, Alaska, the cast aluminum yoke on the nose gear failed and the nosewheel left the airplane. The strut then failed and the aircraft was further damaged.

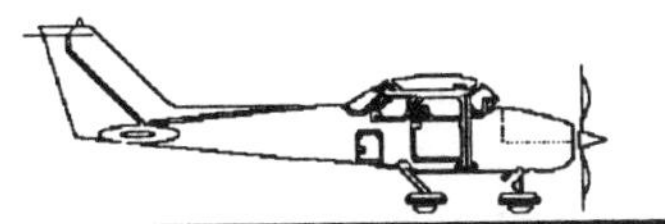
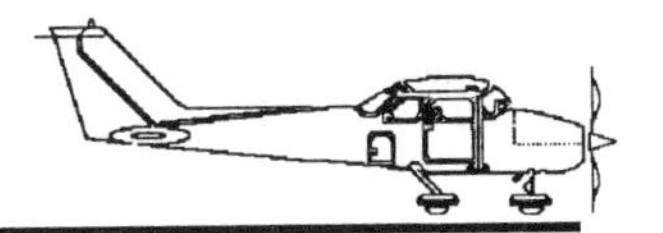

CHAPTER SIX

THE CESSNA 210

How Much 210 Can You Afford?

There were 204 210s advertised in a spring, 1989 issue of "Trade-A-Plane," including 38 pressurized models, 48 normally aspirated and 118 turbo 210s. Their prices were averaged for this analysis.

Looking at the 1980 model year as an example, the normally aspirated asking price was $66,500. The turbo was $79,000, while the pressurized model was $80,000.

The main problem you have to watch for is the skill of the pilot who flew it before you, as was the case with the 182. Many pilots shock cool the high performance engine, and it isn't a problem limited to the 210. The pilots discover, through poor planning, they forgot to come down soon enough, and pull the power back. Do that a few times, and cracks develop in the exhaust system and engine case.

Model years of 1969 and earlier had problems with the landing gear saddles cracking, but that seems to be cured in later years.

The pressurized model had a problem with vacuum pumps, especially in IFR conditions when the pilot would turn on the deicing boots. The pump, already overworked, sometimes failed due to the extra stress of inflating the boots.

Still, the 210 is a favorite of most pilots. The six-place model easily handles five adults and full fuel. On one trip a Centurion loaded in that manner made a two-hour flight from Washington, D.C., to Ocracoke, N.C., and returned, on one tank of gas. Even with headwinds, the return flight was under two-and-a-half hours.

Pilots tend to worry a little about the backseat passengers, though. They are blocked, in an emergency, by two rows of seats.

Some of the ads in the issue of *Trade-A-Plane* had phrases such as, "...has the good, Lycoming 2,000 hr. TBO." Let's do some investigative work on the claim by tracing the history of the airplane.

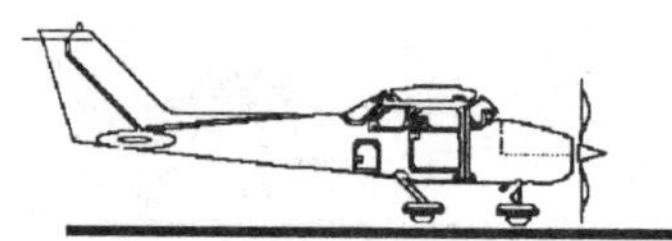
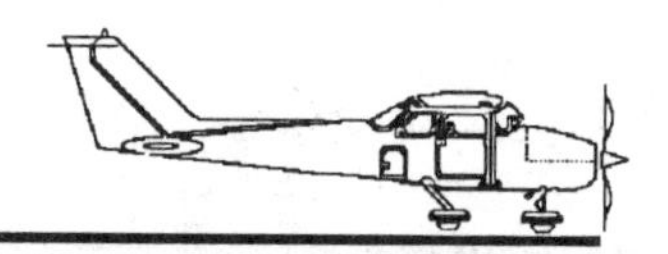

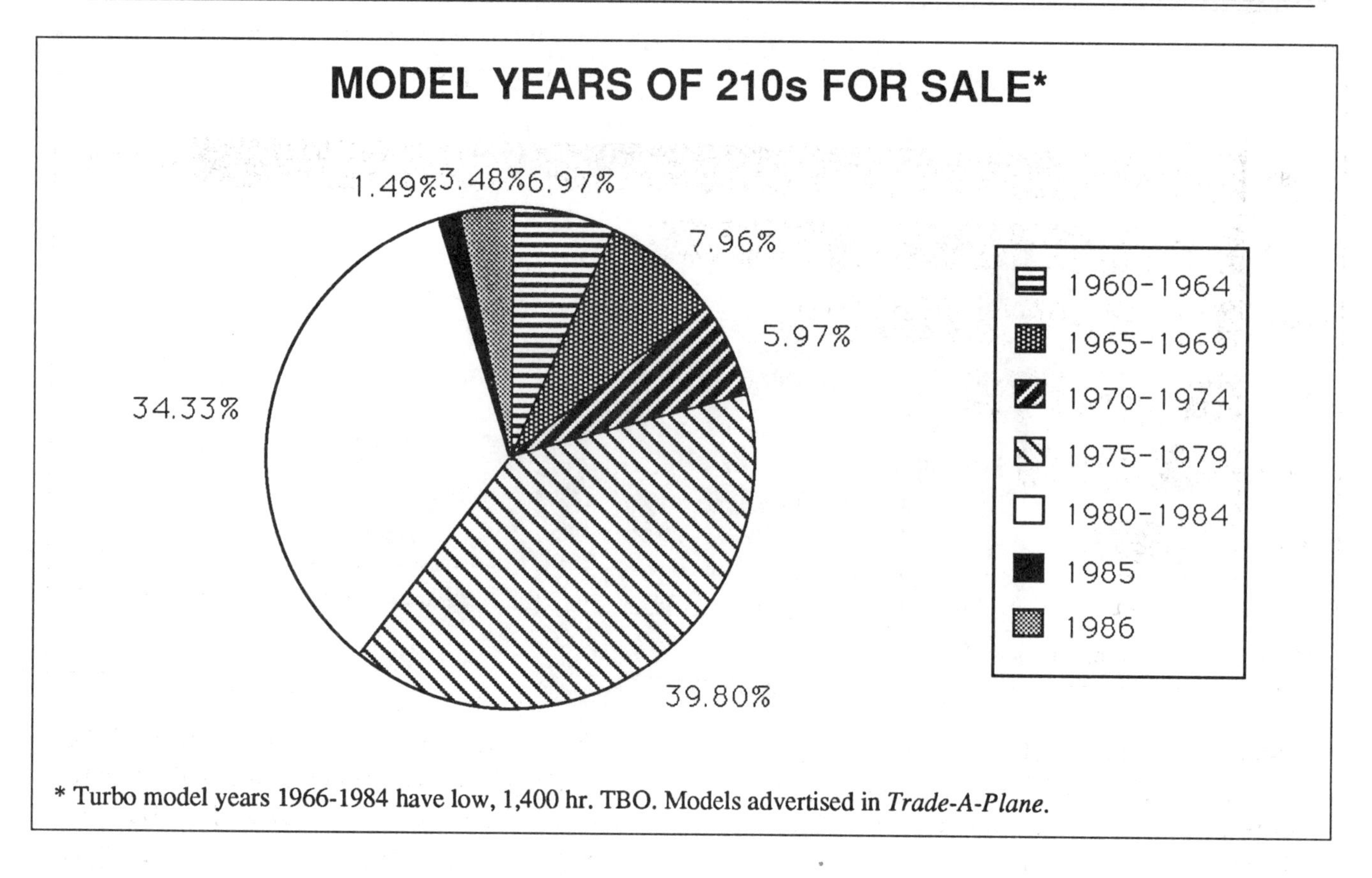

* Turbo model years 1966-1984 have low, 1,400 hr. TBO. Models advertised in *Trade-A-Plane.*

The 210, not called the Centurion at first, began in 1960 as a four-place airplane (the six-place began in 1963) with the IO-470-E Continental engine. It had a TBO of 1,500 hrs. The engine switched to the IO-470-S in 1962, still with a 1,500 TBO.

In 1964 the airplane picked up the name

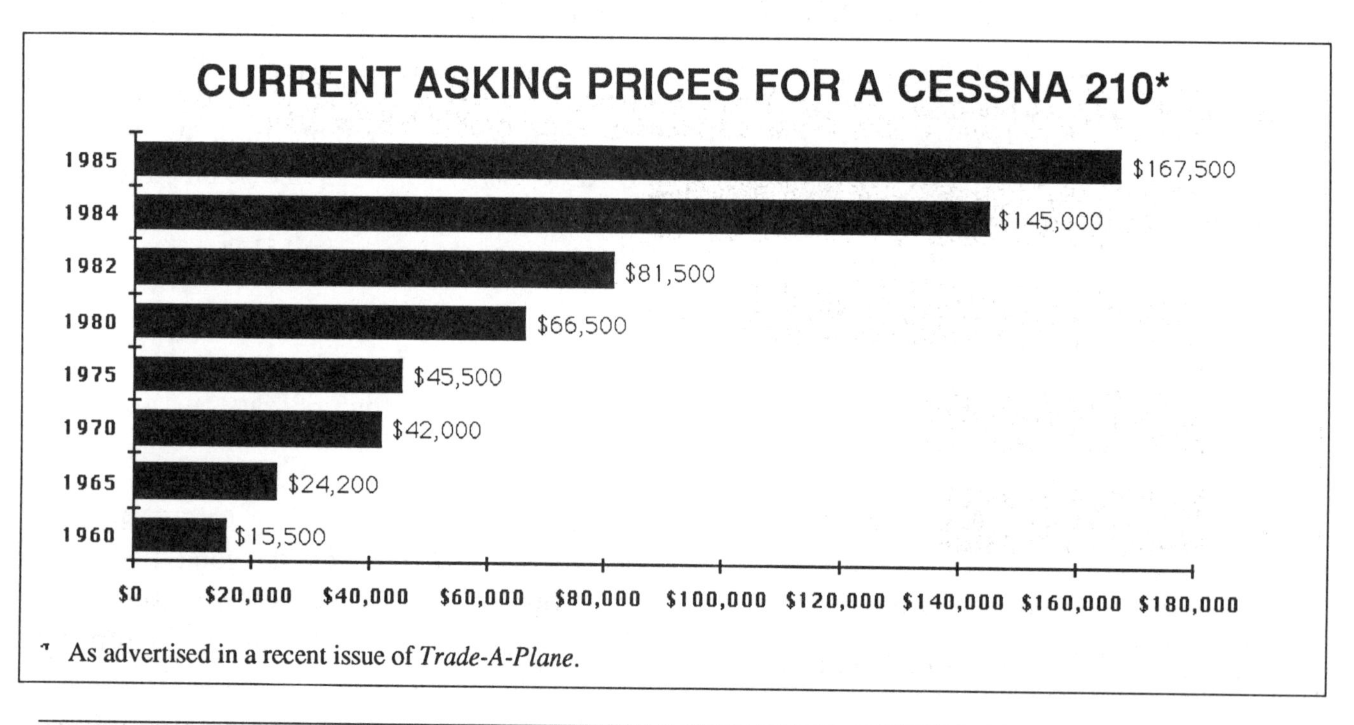

* As advertised in a recent issue of *Trade-A-Plane.*

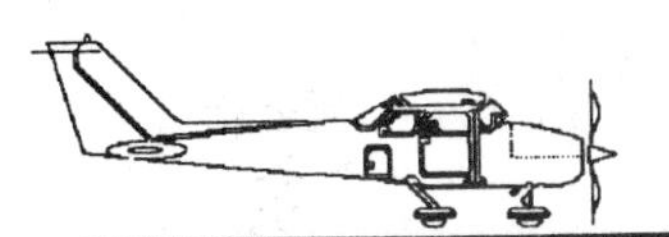
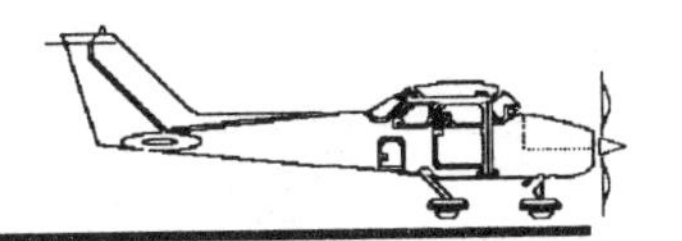

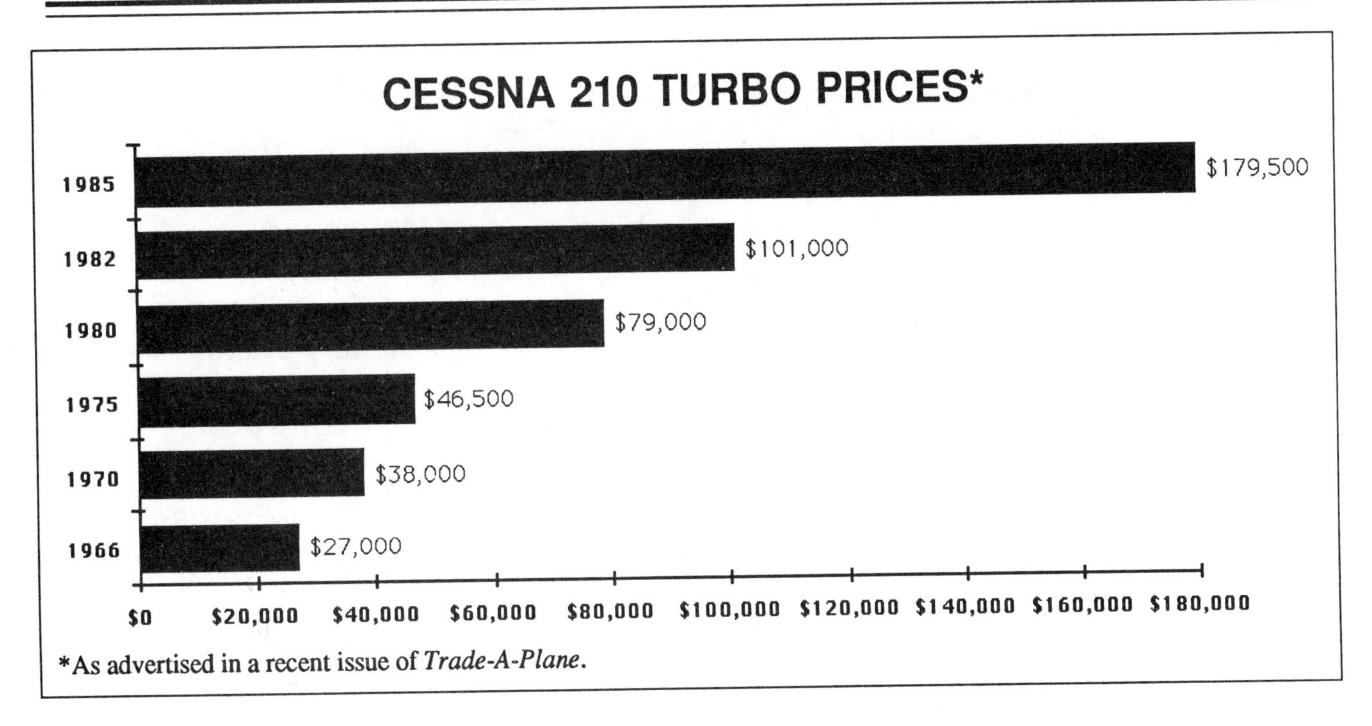

*As advertised in a recent issue of *Trade-A-Plane*.

Centurion and got the IO-520-A engine by Continental with a TBO of 1,700 hrs. (unless you use it in aerial spraying, in which case the TBO is only 1,200 hrs).

In 1966 the first turbo Centurion came along with a TSIO-520-C by Continental that had 1,400 hr. TBO. TBOs did not change again until the 1978 turbo Centurion began wearing the Continental TSIO-520-R with a TBO of 1,600 hrs., assuming certain modifications have been made. If they have not, that engine has only a 1,400 TBO. The engines change a few times after that, but the TBO never got better than 1,600 hrs.

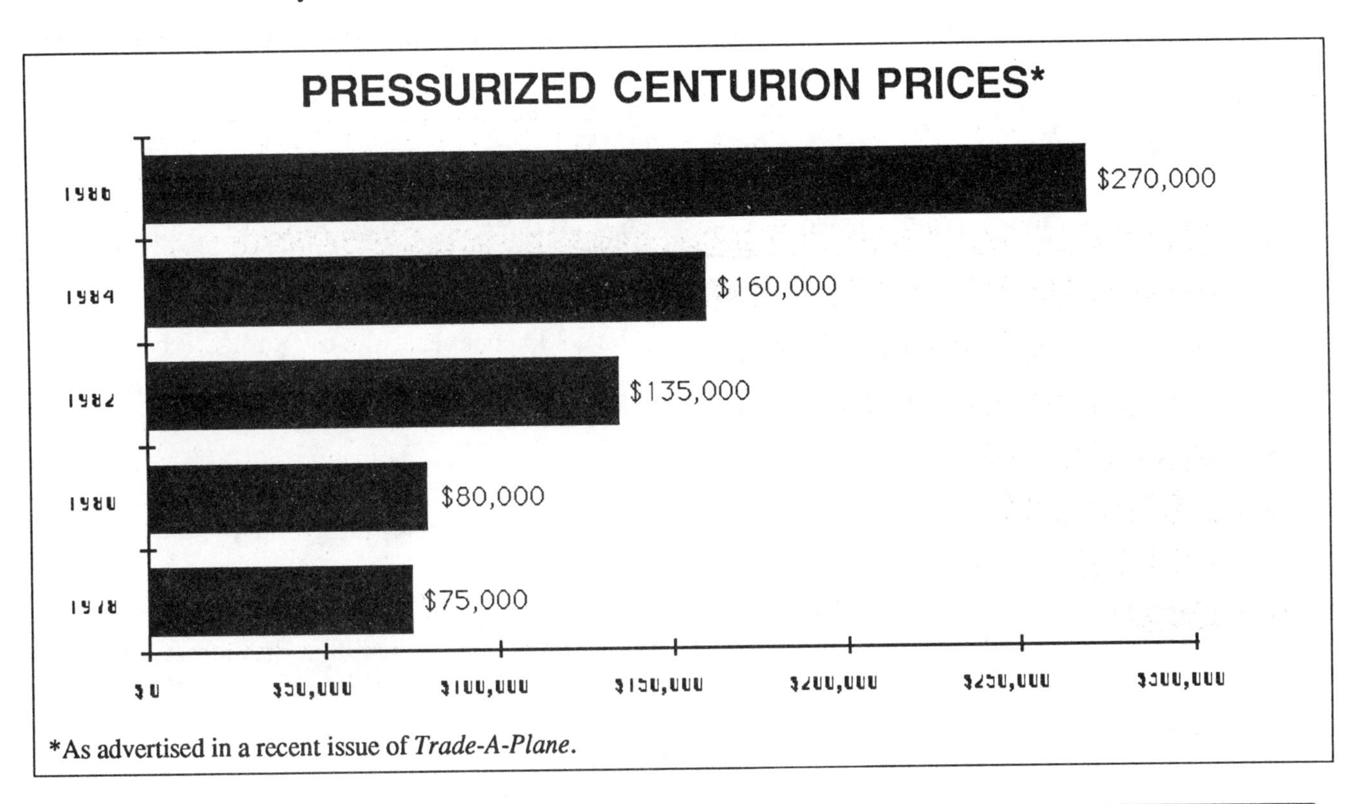

*As advertised in a recent issue of *Trade-A-Plane*.

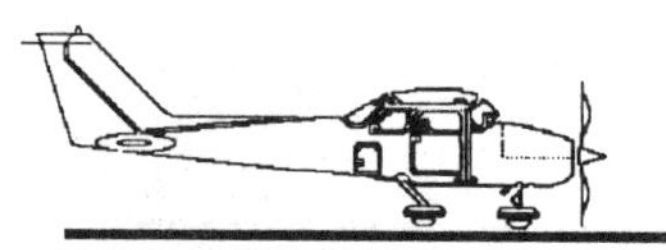

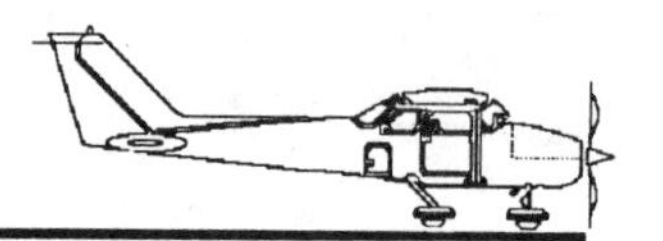

Thus, we have our answer to the claim in the ad. His aircraft has an engine not certificated with the airplane. It has either the TSIO-520-BE or AE, with a TBO of 2,000 hrs. As long as he has a supplemental type certificate granting permission from the FAA for the different engine, his aircraft is legal. Ask to see that certificate.

The pressurized Centurion began in 1978.

The 300 hp. Centurion did not start until 1970.

A Tale Of Two 210s

Let's compare two privately owned 210s from 1986 to 1989, looking at all maintenance costs and annual inspections. They are similar models, and do not have turbochargers or pressurized cabins.

Plane one's history (this is an actual aircraft operating on the East Coast) begins in late 1986 with a $500 bill because the landing gear would not retract. It took $120 in labor to put on a $340 engine-driven hydraulic pump.

The next month, the landing gear would retract but wouldn't remain that way, so it was off to the maintenance shop for another $200 repair.

Three months later it took $80 to tighten some screws and stop a fuel leak at the right wing root. Things went well until four months later when it took $160 to replace the airplane's main battery. The plane was at home in the maintenance shop by now, since it had been there two days earlier to have the magnetic compass replaced by a card compass ($30).

Five months passed, then it was time for an annual inspection which cost $5,600. Ouch! (The Cessna Pilots Association suggested in a phone call this cost is "out of context," and that actual operating costs of a 210 are about $75 an hour. A dealer supports that figure, adding that $75 to $80 is an accurate figure.) The previous inspection had been done by another repair station which set the magneto points too close. It took $130 to reset them properly. Here are some other items of interest from that annual:

> Two tires were needed for $70 and $88. (A tube was $45.)

> A filter cost $266.

> Inspection panels had frozen and corroded screws, requiring lots of labor to remove, costing $100.

> The prop spinner bulkhead broke, costing $600 for a new part and $120 to put the part on.

> The cowl flap was hanging by a thread, its hinges gone and rivets sheared, costing $200 in labor alone to repair.

> Three little things, seals, retainers and rivets, cost three big prices: $130, $115 and $100.

Notice the above inspection could have been over $200 cheaper if the owner had loosened or replaced his own inspection panel screws and had not gone to a poor quality maintenance shop which set magneto points wrong.

210 Number Two

We join the flow of bills for the second 210 at about the same time in 1986. They start with $175 for cleaning an injector nozzle to correct a rough-running engine, and replace a voltage regulator to

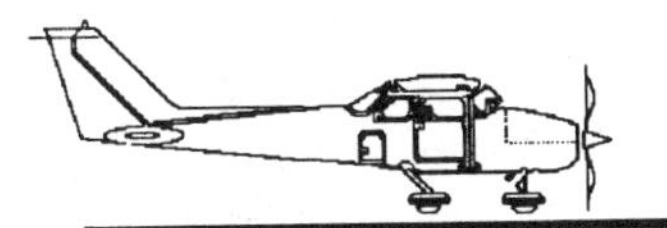
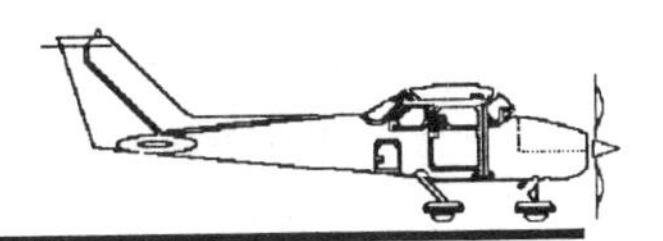

correct a high voltage light.

Four months later, still in 1986, it took $375 to correct a magneto problem. Fourteen days later it took $265 to put in a new gear-down switch. The old one was popping the gear circuit breaker.

Two months later it was time to pay $670 for a $310 vacuum pump and an oil and filter change. The pump cost $115 in labor to put on the airplane. Labor for the oil change was $65.

Two months after that it was time for an annual that cost only $2,000. Some of the things found wrong in that annual include:

> Baffle cracked on a cylinder — $60.

> ELT battery — $35.

> Cylinder intake pipe drain line broken — $50.

> An eight-inch crack in the leading edge of the right flap — $50.

> Wire off the panel for the gear actuator — $60.

Most of those costs are just the labor alone.

The next month (can't you hear the owner saying, "But I just had it annualed!") it cost $550 to fix the right vacuum pump. The left one (see above) had failed four months earlier and cost $100 more to fix.

Then, all was quiet for seven months. The airplane purred along as advertised. But in mid-

Sample Yearly Insurance Rates For A Cessna 210*

Year	Value	Pvt. 25 rg** 250 hr.	Pvt. 150 rg** 500 hr.	Pvt. 500 rg** 1,300 hr.
1960 (4 seats)	$15,500	$1,950	$1,800	$1,525
1970 (6 seats)	42,000	3,200	2,935	2,500
1975	45,500	3,350	3,050	2,525
1980	66,500	4,200	3,825	3,200
1982	81,500	4,800	4,381	3,660
1984	145,000	N/A	6,660	5,500
1985	167,500	N/A	7,300	6,078
1986	270,000	N/A	9,696	8,070

* No difference in premium for turbo 210.
40% increase for pressurized 210.
The liability limits are based on $500,000 combined single limit (includes bodily injury to the public, property damage liability and passenger liability limited to $50,000 each person).

** RG refers to retractable gear experience.

1987 it took $600 for a 50-hour inspection to see why there was a 200 rpm. drop on a magneto, and to fix the left strobe light (bulb price: $80).

The next month the fuel flow transducer broke, costing $60.

The plane flew on into 1988, six months since the transducer broke, and had another vacuum pump failure. It was under warranty, but the labor to replace the pump was not free and cost $180.

Then it was time for another annual — a $7,000 annual! However, part of that was caused by an airworthiness directive. Labor alone was $3,300.

The biggie for this annual was a $2,700 prop overhaul caused by an airworthiness directive. (This was an older model 210.) The labor to put the prop on and take it off was $130. Here are some other problems found in that annual:

> It took $600 to check the oxygen system.

> It took $260 to fix the left strobe light.

> The cabin heat shroud was checked for $110.

> Cracks in the engine mount heat deflector and the cylinder intake hose heat deflector cost $65 and $85 respectively.

> The left drain for the main fuel tank was leaking, and it took $280 to stop it.

> The main gear door weather stripping was damaged, costing $65.

> An overhead console support bracket cost $70 to repair.

> Wing inspection panels cost $60, while stripped nut plates on floor inspection panels cost $65.

The song, "You Gotta Have Heart," should read, "You Gotta Have Dough" for 210 owners. Again, it should be noted that most get by with $75-80 an hour. But make sure all ADs have been done so you are not surprised by a $2,700 prop AD.

Cessna 210 Type Certificate

The Cessna 210 type certificate has one of those notes the FAA warns used plane buyers about. The P210N and R windshield, rear cabin top windows, side windows and ice detector light lens must be "retired" (thrown out) at 13,000 hours. Obviously that is not much of a problem. There are few 13,000 hr. P210N or P210R airplanes for sale. But it is nice to know as a potential buyer.

The other notes pertain to placards, similar to most of the notes for the 152, 172 and 182. Your mechanic can check the placards for you.

Not only did the aircraft change engines and props as the model years passed, the 210 grew from four people to six. If you have rented a 210, mostly likely it was a six-passenger machine, and you might think they are all like that. So, the type certificate is necessary reading.

CESSNA 210 (Model Year 1960). Four place.
Continental IO-470-E, 2,625 rpm., (260 hp.). 100/130 grade gas, 65 gal. (55 gal. usable). Maximum weight 2,900 lb. Propeller, Hartzell HC-A2XF-1/8433-2, McCauley D2A36C33/90M-8 or D2A34C49/90A-8 or D2A34C58/

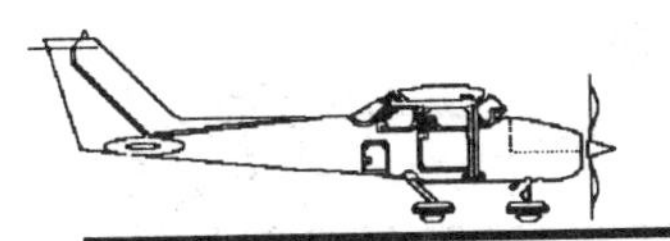
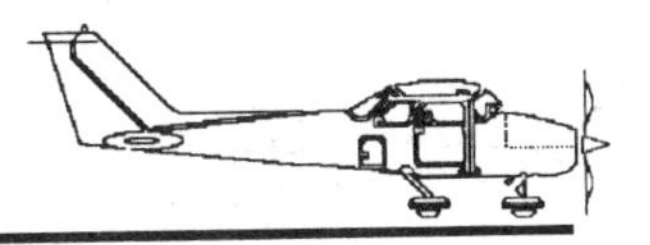

90AT-8. Max cruise, 152 knots.

CESSNA 210A (1961). Four place.

Continental IO-470-E, 2,625 rpm. (260 hp.). 100/130 grade gas, 65 gal. (55 gal. usable). Maximum weight, 2,900 lb. Propellers same as 210 above. Max cruise, 152 knots.

CESSNA 210B (1962), 210C (1963). Four place.

Continental IO-470-S, 2,625 rpm., (260 hp.). 100/130 grade gas, 65 gal. **(63.4 usable).** Maximum weight, **3,000 lb.**. Propellers same as above. Max cruise, **165 knots.**

CESSNA 210-5 (205) (1963), 210-5A (205A) (1964). Six place.

Continental IO-470-S, 2,625 rpm., (260 hp.) 100/130 grade gas, 65 gal., (63.4 usable). Maximum weight, **3,300 lb.**. Propellers: Hartzell HC-A2XF-1A13.5/8433-2, McCauley D2A36C33/90M-8 or D2A34C49/90A-8 or D2A34C58/90AT-8. Max cruise, 148 knots.

CESSNA 210D (1964). Four place.

Continental IO-520-A, 2,700 rpm. (285 hp.). 100/130 grade gas, 65 gal. **(63.4 gal. usable).** Maximum weight, **3,100 lb.** Propeller, McCauley D2A34C58/90AT-8. Max cruise, 165 knots.

CESSNA 210E (1965). Four place.

Continental IO-520-A, 2,700 rpm. (285 hp.). 100/130 grade gas, 65 gal. **(63.4 usable).** Maximum weight, **3,100 lb.** Propellers: McCauley E2A34C64/90AT-8, McCauley E2A34C73/90AT-8. Max cruise, 165 knots.

CESSNA T210F (1966). Four place.

Continental TSIO-520-C, 2,700 rpm., 32.5 in. hg. (285 hp.). 100/130 grade gas, 65 gal. (63 gal. usable). Maximum weight, **3,300 lb.**. Propellers: McCauley E2A34C70/90AT-8, McCauley D3A32C77/82NK-2, McCauley D3A32C88/82NC-2. Max cruise, 165 knots.

CESSNA 210F (1966). Four place.

Continental IO-520-A, 2,700 rpm., (285 hp.). 100/130 grade gas, 65 gal. **(63 ga. usable).** Maximum weight, **3,300 lb**. Propellers: McCauley E2A34C73/90AT-8, McCauley D3A32C77/82NK-2, McCauley D3A32C88/82NC-2. Max cruise, 165 knots.

CESSNA T210G (1967 — four place), T210H (1968 — four place).

Continental TSIO-520-C, 2,700 rpm. 32.5 in. hg. (285 hp.). 100/130 grade gas, **90 gal. (89 usable). Maximum weight, 3,400 lb.** Propellers: McCauley E2A34C70/90AT-8, McCauley D3A32C88/82NC-2, McCauley D3A32C77/82NK-2 (T-210G only). Max cruise, 165 knots.

CESSNA 210G (1967) — four place), 210H (1968 — four place).

Continental IO-520-A, 2,700 rpm. (285 hp.). 100/130 grade gas, 90 gal. (89 usable). Maximum weight, 3,400 lb. Propellers: McCauley E2A34C73/90AT-8, McCauley D3A32C88/82NC-2. Max cruise, 165 knots.

CESSNA T210J (1969) Four place.

Continental TSIO-520-H, 2,700 rpm., 32.5 in. hg. (285 hp.). 100/130 grade gas, 90 gal. **(89 usable).** Maximum weight, **3,400 lb.** Propellers: McCauley E2A34C70/90AT-8, McCauley D3A32C88/82NC-2. Max cruise, 165 knots.

CESSNA 210J (1969). Four place.

Continental IO-520-J, 2,700 rpm., (285 hp.). 100/130 grade gas, 90 gal. **(89 gal. usable).** Maximum weight, **3,400 lb.** Propellers: McCauley E2A34C73/90AT-8, McCauley D3A32C88/82NC-2. Max cruise, 165 knots.

CESSNA 210K/T210K (1970 — six place),

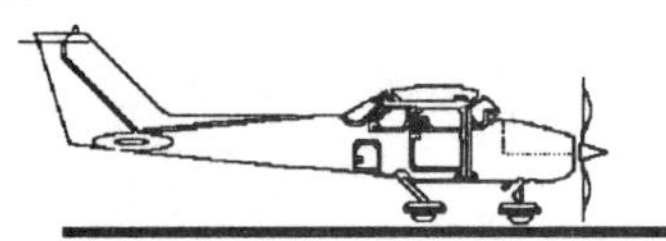
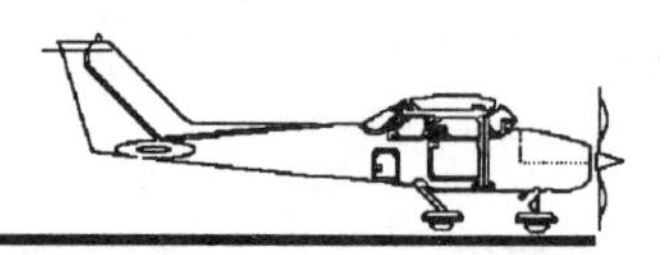

210L/T210L — six place.

Continental IO-520-L, five minutes at 2,850 rpm., then 2,700 rpm. **(300 hp.)** Model T210K/T210L. Continental TSIO-520-H, 2,700 rpm., 32.5 in. hg. (285 hp.). 100/130 grade gas, 90 gal. (89 usable). Maximum weight, **3,800 lb.** Propellers: Model T210K/T210L, McCauley E2A34C70/90AT-8, McCauley D3A32C88/82NC-2. Max cruise, 210K/T210K, 210L/T210L, 168 knots indicated.

CESSNA 210M/T210M (1977). Six place.

Engines: model 210M, Continental IO-520-L, five min. at 2,850 rpm. (300 hp.), then 2,700 rpm. (285 hp.). Model T210M, Continental TSIO-520-R, five min. at 2,700 rpm., 36.5 in. hg. (310 hp.), then 2,600 rpm. (285 hp.). Fuel: 210M/T210M, 100/130 grade gas for serial numbers 21061574 through 21062273, 210M/T210M, 100LL/100 for serial numbers 21062274 through 21062953. 90 gal., (89 gal. usable). Maximum weight, 3,800 lb. Propellers: model 210M (serial number 21061574 through 21062273), McCauley D3A32C88/82NC-2; model 210M (serial numbers 21062274 and up), McCauley D3A34C404/80VA-0. Propellers for T210M, McCauley D3A34C402/90DFA-10. Max cruise: 210M, 168 knots; T210M, 165 knots.

CESSNA P210N (1978). Six place.

Engine: serial numbers P21000001 through P21000760, Continental TSIO-520-P, 2,700 rpm. for five min. (310 hp.), then 2,600 hp. (285 hp.); serial numbers P21000761 and up, Continental TSIO-520-AF, five minutes at 2,700 rpm.(310 hp.), then 2,600 rpm. (285 hp.). 100LL/100 grade gas, 90 gal. (89 gal. usable for P21000001 through P21000760, 87 gal. usable for higher serial numbers). Maximum weight, **4,000 lb. takeoff, 3,800 lb. landing, 4,016 on the ramp.** Propeller, McCauley D3A34C402/90DFA-10. Max cruise, 167 knots indicated.

CESSNA 210N/T210N (1979). Six place.

Continental IO-520-L, 2,850 rpm. for five minutes (300 hp.), then 2,700 rpm. (285 hp.). 100LL/100 grade gas, 90 gal. (89 gal. usable, 87 gal. usable for serial numbers 21064536 and up). Maximum weight, 3,800 lb., 3,812 on the ramp. Propeller, D3A34C404/80VA-0. Max cruise, 165 knots.

MODEL T210N Continental TSIO-520-R, 2,700 rpm. for five minutes (310 hp.), then 2,600 rpm. (285 hp.). 100LL/100 grade gas, 90 gal. (89 gal. usable, 87 gal. usable for serial numbers 21064536 and up). Maximum weight, 4,000 lb. takeoff, 3,800 lb. landing, 4,016 on the ramp. Propellers: McCauley D3A34C402/90DFA-10. Max cruise, 168 knots.

CESSNA P210R, Pressurized Centurion (1985). Six place.

Continental TSIO-520-CE, 2,700 rpm., 37 in. hg. **(325 hp.).** 100LL/100 grade gas, 90 gal. (87 gal. usable), **long range 120 gal. (115 gal. usable).** Maximum altitude, 25,000 ft. Maximum weight, 4,100 lb. takeoff, 3,900 lb. landing, **4,116 lb. on the ramp.** Propeller, McCauley D3A36C410/80VMB-0. Max cruise, 167 knots.

CESSNA T210R (1985 — six place), 210R — six place.

Continental IO-520-L, five minutes at 2,850 rpm. (300 hp.), then 2,700 rpm. (285 hp.). 100LL/100 grade gas, 90 gal. (87 gal. usable), long range 120 gal. (115 gal. usable). Maximum weight, 3,850 lb., 3,862 lb. on the ramp. Propeller, McCauley D3A34C404/80VA-0. Max cruise, 167 knots.

MODEL T210R Continental TSIO-520-CE, 2,700 rpm., 37 in. hg. (325 hp.) 100LL/100 grade gas, 90 gal. (87 gal. usable), long range 120 gal. (115 gal. usable). Maximum weight, 4,100 lb.

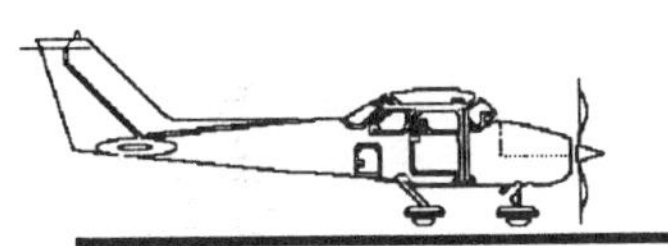

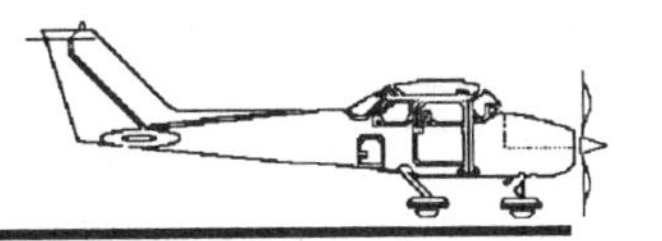

takeoff, 3,900 lb. landing, 4,116 lb. on the ramp. Propeller, McCauley D3A36C410/80VMB-0. Max cruise, 167 knots.

A Talk With A Dealer

Porsche Galesburg, Peoria, Ill., gained fame for putting the first Porsche engine on a Cessna 182 in the U.S. When Porsche Galesburg president Harrel Timmons was interviewed for this book, the firm was just completing certification of a Porsche engine on a 172.

However, there is more to the firm than just Porsche engines. Its active sales department has sold many Cessna 210s. "We sell eight turbos for every two normally aspirated 210s," said Timmons, who jokingly refers to himself as head line boy.

A recent check of airworthiness directives for 1986 turned up more for the turbo and pressurized versions than the normally aspirated 210s. However, Timmons said most of the ADs are "pretty insignificant."

"The turbo is the best protection against icing. It is much quieter and has outstanding performance," he said. The normally aspirated 210 runs out of steam at 9,000 ft., he said.

He has sold about a hundred 210s in past years, and operates three in an air taxi service.

"Most of them were owned. I am not keen on brokered airplanes," he said. "The surprises always come back to haunt you. You need to own (brokers do not own the airplanes they sell) the plane to do the things required to get it ready for sale." He sends the props out to be overhauled, rather than doing the work in shop, and normally pays between $1,500 and $3,000. Owners are supposed to overhaul the prop every time they overhaul the engine, which in the 210 is every 1,400 hrs. to 1,700 hrs. There is no firm overhaul requirement, however.

Timmons does engine and airframe work in his own shop prior to sale. He was asked if he would pay for all repairs if a buyer offered to pay for a 100 hr. inspection. Timmons said he would, and routinely does that anyway.

He was asked for tips to potential buyers, and Timmons had these suggestions.

"If it is the buyer's first heavy airplane, and he is not accustomed to that type of airplane, he should be aware it does not operate cheaply. He must be very aware of the kind of maintenance it has had. If a whole bunch of maintenance has been let go, even if the numbers in the logbooks look good, then repairs will be expensive. About $10,000 to $17,000 later he finds out he has an expensive airplane.

"For example, I operate three 210s as an air taxi. The exhaust system problems must be fixed when they are still a small problem. Our airplanes get 50 hr. and 100 hr. inspections and operate at $75 to $80 an hour, depending on the airplane. Service bulletins that might be optional to a private owner are not optional to us."

So Porsche Galesburg 210s get lots of routine maintenance, and problems are fixed early. That seems to be the secret of keeping maintenance costs low. Timmons was asked whether cracked landing gear saddles, found often in early service reports on the 210, are still a problem.

"That was on 1969 and older 210s. It hasn't occurred in the later models," he said.

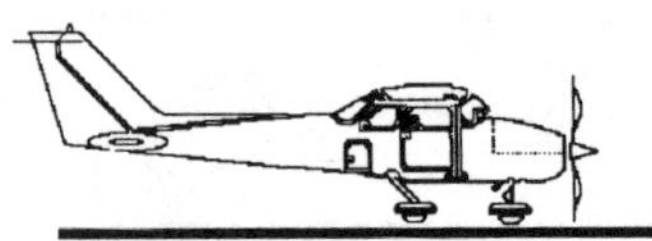
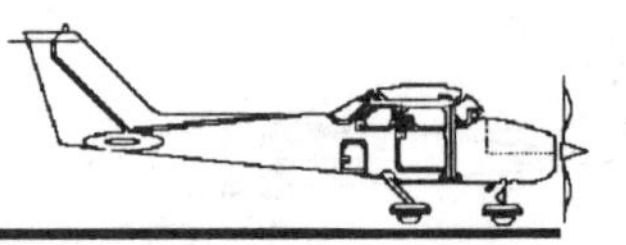

"The most common problem today is with exhaust system and cylinder case cracking. And 80% of those problems are caused by the operator. Sometime they run it too hot, but shock cooling is the biggest bugaboo of 210s. It is so easy to shock cool a 210; a pilot at 8,000 ft. will discover he is 10 miles from the airport and pull the power back.

"We climb at speeds at the top of the white arc as a standard procedure, about 20 knots above the best climb speed.

"A 210 has to be treated as a high performance racing engine. I've run 210s over 2,000 hrs. without touching the engine."

Like several other dealers interviewed for this book, Timmons said most of his sales are going overseas at present, due to devaluation of the dollar in Europe. For Europeans, it is as though there is a 35%-off sale on the U.S. general aviation fleet.

"About 90% of our sales since January [1988] — about a dozen planes — have gone overseas. I have a friend who has sent 14 overseas. During that time, I sold one in the U.S.," Timmons said.

Timmons also revealed that U.S. aviation officials may be unaware of the outflow of airplanes to Europe, because 30% of his planes sold are not deregistered.

"They can operate U.S. airplanes on a U.S. license. IFR privileges are far more lenient," Timmons said.

"It's a real strange situation. There will be no clean, low-time airplanes in the U.S. in three years," Timmons predicted.

Timmons said overhaul costs of a 210 engine range from $15,000 to $25,000.

Centurion Airworthiness Directives

Below is a summary of some of the airworthiness directives for the 210 Centurion, turbo 210 and pressurized 210. It was taken from the FAA's "Book Two," starting in 1971, and is not an attempt to provide a complete list of ADs. Such a list would be quickly out of date.

The idea is to give you a brief picture of the airplane's troubles over the years. By looking at each one, the prospective buyer can tell whether those troubles are big or small.

210 ADs Common to Other Cessna Models

76-04-03.

Cessna 210K, L, T210K, L. To preclude restriction of control movement due to jamming of the ARC PA-500A actuator gear train.

77-12-08.

Cessna 210L, T210L, equipped with electrical ground power receptacles. To prevent unwanted propeller rotation.

79-08-03.

Cessna 210, T210. To prevent in-flight electrical system failure, smoke in the cockpit and/or fire in the wire bundle behind the instrument panel.

79-10-14.

Cessna 210, T210. To provide an alternative source of fuel tank venting in case of fuel tank vent obstruction by foreign material and/or sticking of the fuel tank vent valve.

79-25-07.

Cessna 210, T210. To preclude the possibility

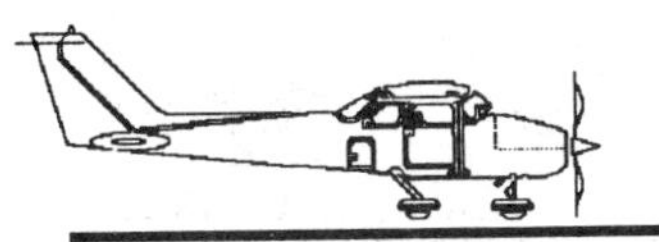

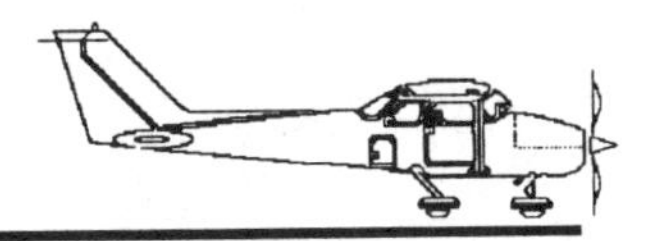

of electrical or electronic component damage or an in-flight fire due to a short between an ungrounded alternator and flammable fluid-carrying lines.

84-10-01.

Cessna 210, T210. To prevent power loss or engine stoppage due to water contamination of the fuel system, install quick drains and check for fuel leakage.

86-19-11.

Cessna 210B through R, T210F-R, P210N, R equipped with fuel reservoirs. To eliminate the possibility of engine power reduction due to contaminated fuel, install quick drains or attach information to the airplane documents.

86-24-07.

Cessna 210C-M, T210K-M, T210F-J, P210N. To prevent engine power interruption due to loss of attachment of the engine controls.

86-26-04.

Cessna 210, many models. To prevent slippage of the pilot/co-pilot shoulder harness.

ADs Unique to the 210

76-04-01.

Cessna 210 to 210D with Electrol main gear rotary actuator assemblies. To decrease the possibility of main gear extension failure.

76-14-07.

Cessna 210 through 210J, T210F through T210J. To decrease the possibility of main landing gear extension failures, due to cracked saddles.

77-16-05.

Cessna 210, T210. To prevent malfunction of the fuel selector valve.

78-07-01.

Cessna T210. To preclude engine oil pump failure due to contamination by the turbocharger thrust bearing anti-rotation pins and failure of the turbocharger shaft.

78-11-05.

Cessna 210M, T210M, P210. To preclude increased flight control forces caused by an autopilot actuator that has failed to disengage when the autopilot is disconnected.

78-26-12.

Cessna 210G, H, J, M, N, T210G, H, J, M, N, P210N. To detect binding of fuel quantity transmitter float arm and assure proper operation of the fuel quantity indicating system.

79-03-03.

Cessna T210M. To preclude failure of engine oil pressure loss caused by turbocharger oil scavenge pump ingestion of failed turbocharger thrust bearing anti-rotation pins.

79-15-01.

Cessna 210, T210, certain serial numbers. To provide instructions for recognition of fuel system vapor blockage and operating procedures to restore normal fuel flow.

79-19-06.

Cessna 210L, T210L, 210M, T210M and P210N modified to incorporate Symbolic Displays fuel flow indicating system per STC SA3835WE. To prevent a possible fuel leak caused by installation of Symbolic Displays, Inc., fuel flow indicating system.

80-04-09.

Cessna T210M, N and P210N. To preclude

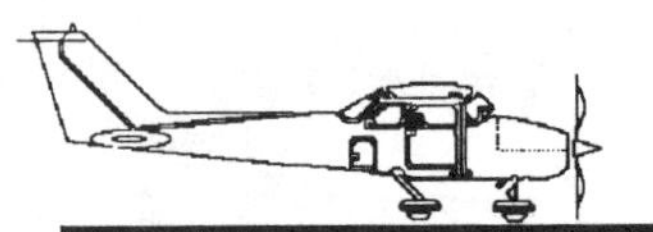
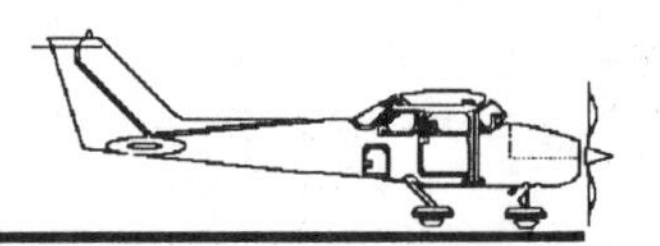

engine power interruptions and fuel flow fluctuations, modify fuel system.

80-07-01.

Cessna T210. To preclude failure of the engine oil pressure and scavenge pump drive shaft and resulting oil pressure loss caused by turbocharger oil scavenge pump ingestion of failed turbocharger thrust bearing anti-rotation pins.

80-21-03.

Cessna 210K, L, M, N, T210K, L, M, N and P210N which have the King Radio KFC-200 autopilot with the roll axis servo actuator mounted in the right wing. To prevent loss of roll axis flight control.

81-23-03.

Cessna P210N, certain serial numbers, with 25 or more hours in service. To ensure the integrity of the engine exhaust system, inspect for cracks.

82-06-10.

Cessna 210L/T210L, 210M/T210M, 210N/T210N, P210N equipped with the Airborne model 442CW-8 vacuum pump, pneumatic deicer boots and vacuum-operated altitude indicator. To prevent loss of vacuum-driven attitude instruments resulting from failure of the single Airborne vacuum pump during flight in IFR conditions. (Failure generally occurred when deicer boots were activated.)

85-02-07.

Cessna 210, T210, several model years, and P210N. To eliminate the possibility of loss of the fuel selector roll pin installation.

85-03-01.

Cessna 210, T210, several model years. To preclude the possibility of engine controls failure and loss of engine power control.

85-10-02.

Cessna 210, several model years, equipped with the Continental IO-520 engine. To eliminate the possibility of engine power reduction due to ingestion of pieces of a failed engine induction airbox outboard duct due to lower skin cracks.

85-11-07.

Cessna P210N, R, T210R. To prevent possible separation of the turbocharger oil reservoir outlet fitting and subsequent rapid loss of engine lubricating oil.

Service Bulletins For The 210

Here is a sampling of service bulletins over the past decade:

(SEB 86-3) cylinder barrel inspection, (SEB 86-2) altimeter inspection and modification, (SEB 85-19) hydraulic power pack drain line, (SEB 85-18) standby electric vacuum pump, (SEB 85-16) fuel tank vent improvement, (SEB 85-14) pilot operating handbook for 1985 models, (SEB 85-11) turbo reservoir inspection, (SEB 85-7) elevator and trim tab inspection, (SEB 85-5) elevator trim tab actuator attach bracket inspection, (SEB 85-2) fuel quantity transmitter gasket sealing improvement, (SEB 85-1) pilot operating handbook revisions — 1984 models, (SE 84-23) low vacuum warning light installation, (SE 84-21) nose wheel shimmy troubleshooting, (SE 84-20) R1 induction airbox improvement, (SE 84-19) instrument light dimming circuit troubleshooting, (SE 84-17) horizontal stabilizer attach bracket improvement, (SE 84-9) **rubber fuel cell drain valve relocation, (SE 84-8) fuel reservoir tank quick drain valve installation, (SE 84-4) rubber fuel cell inspection.**

In 1983 there was an interesting service bulle-

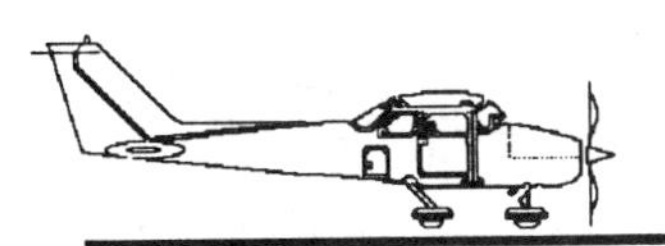
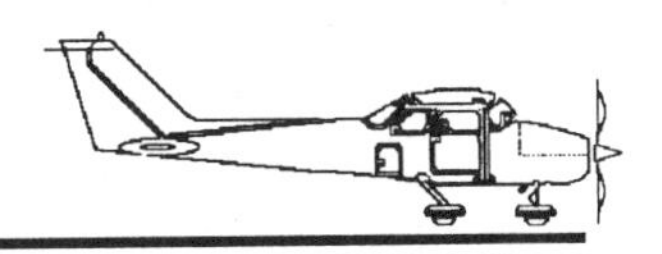

tin, (SE 83-28), titled “special wing main spar center lug inspection.”

In 1980 there were six that appear to be more serious than the rest: (SE 80-95) fuel pump through bolt torque inspection, (SE 80-93) crankshaft oil seal retainer TCMTSIO 520 M/P and R engines, (SE 80-68) 400B autopilot/integrated flight control system, (SE 80-41) Slick magneto impulse couplings, (SE 80-33) freezing of fuel pump/gear warning, and (SE 80-37) alternator regulator case sealing.

The gear warning horn and fuel boost pump actuator was also the subject of a bulletin in 1977 (SE 77-20). Also, in that year, there was a fuel tank inspection (SE 78-10) and a bulletin about the number one avionics switch guard (SE 77-34).

Although there was no specific airworthiness directive or service bulletin, the 210 has been known to suffer wing stress damage in the hands of the wrong pilot. It requires a lot of pilot skill and experience to fly a heavy airplane. Sometimes a pilot will use the “ton of bricks” approach, dropping in from too high with throttle retarded—just the way it was done in the 152 during private pilot training. The newer 210s have no wing struts, and all those people and all that fuel makes for a rock hard landing. Check the log books for wing overstress repairs.

Analysis Of 210 Accidents For 1986

The friendly young man at the National Transportation Safety Board in Washington, D.C., greeted a visitor with, “Boy! You picked all the aircraft with the worst records.” The visitor had come to pick up a computer printout for this book of mechanical failures for the Cessna 150/152, 182 and 210.

What had surprised him was the large number of 210 accidents, over 2,000 since NTSB began computerized records in 1964. Once again, the ability of NTSB records to give an aircraft an undeserved reputation was at work. Sure, the number sounds high, but a look at the individual years reveals 92 accidents or incidents per year. The Cessna 172 has several years where it the total hits 140, and it is considered a safe, easy airplane to fly.

As in all the other models examined, records show it is the pilot, not the aircraft, that is at fault the majority of the time. For the 210 in 1986, only 13 accidents were mechanical — the fault of the airplane. Yes, we wish the total weren’t the unlucky number 13, but luckily it is a small number.

Let’s take a look:

Feb. 17, 1986

An oil passage became blocked on a P210N. The passage supplied oil to the number one connecting rod bearing, causing the rod to fail. The engine quit and the plane made a forced landing on unsuitable terrain, resulting in three fatalities and two serious injuries at Dulce, N.M.

Feb. 27

The number two main bearing “moved” on a Cessna 210M, causing a bending overload of the crankshaft and separation of the crankshaft at the number three cheek. The bearing disintegrated and the engine failed while on an ILS approach at Pontiac, Mich. The aircraft landed on a frozen lake and there were no injuries.

March 12

Three wires were loosely connected to the alternator circuit breaker of a T210N during a flight at White Plains, N.Y. Approximately one and a half turns of the screws were required to secure the wires to the terminal. The pilot noticed

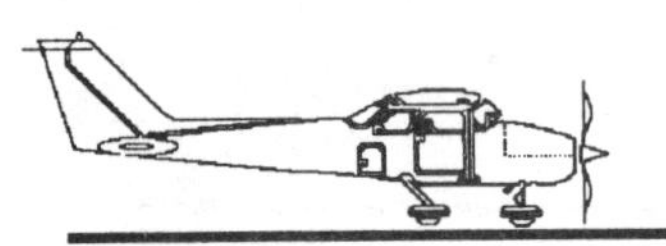

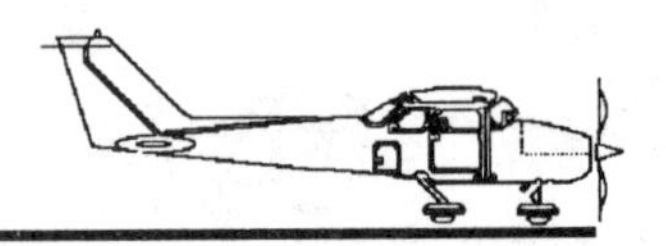

a discharge on the ammeter and an open alternator circuit breaker after ATC reported loss of transponder signal. He lowered flaps and gear electrically, but the nose gear collapsed and the main gear were still in their bays.

March 25

A Cessna 210E at Jacksonville, Fla. could not get an indication of gear up. Another aircraft flew under the aircraft and reported the nose gear and left main gear were up but the right main gear was partially extended. The aircraft was landed in this configuration by the 63-year-old commercial pilot. A secondary relief valve in the hydraulic power pack was being held open by contamination, allowing hydraulic pressure to bypass the landing gear actuators. The logbook showed no evidence the hydraulic power pack had ever been overhauled or that a modification to remove the secondary relief valve had been complied with.

May 3

A P210N experienced a power loss 10 miles south of Long Beach, Calif., Airport and landed on a golf course one-quarter mile from the airport. The number two piston has melted due to extreme high temperature distress, and the connecting rod fractured in overload. The engine had been installed at Cincinnati, Ohio, on May 2. Metallurgical examination disclosed misalignment of the connecting rod and piston/cylinder barrel.

July 25

A T210 made an emergency landing on a road, clipping a fence, after the number five rod bolt failed, with no injuries to the four passengers at Battle Mountain, Nev.

July 30

A 210B made a forced landing and struck trees near Hopewell, Fla., after the crankshaft failed due to fatigue cracking between the numbers one and two rod bearing journal. The engine was not disassembled and inspected as required, engine records revealed, during a sudden stoppage in mud 25 hours prior to the accident. Additionally, it had 1,788 total flight hours since factory overhaul, although the TBO for that engine (Continental IO-470-S) is 1,600 hours.

Aug. 3

The right master cylinder of a T210 stuck, causing the nose gear to cock to the right, and it veered off the runway to the right during a landing roll.

Aug. 13

The engine crankshaft failed on a P210N at 10,500 ft. The aircraft was first vectored to a sod strip that had been closed and was plowed, then was vectored to another airport but did not have sufficient altitude to glide that far. The aircraft landed in a plowed field with serious injuries to two.

Sept. 14

A commercial pilot with an overdue medical certificate in a T210M which was two months overdue for an annual inspection experienced a fuel pressure failure on takeoff at McAlester, Okla. A functional check of the fuel pump revealed no deficiencies and there was normal engine operation at 1,000 rpm., but the runup had to be terminated when a fuel leak developed at the mixture control shaft. One serious and two minor injuries.

Sept. 18

A P210 near Pasco, Wash., experienced a catastrophic engine failure with a flash fire at the turbocharger. The number three piston, connecting rod and cap had separated in flight. A top overhaul had been performed on the engine 11 months (163 flight hours) prior to the accident. Although the pilot undershot the emergency landing, catching the landing gear on a fence, there

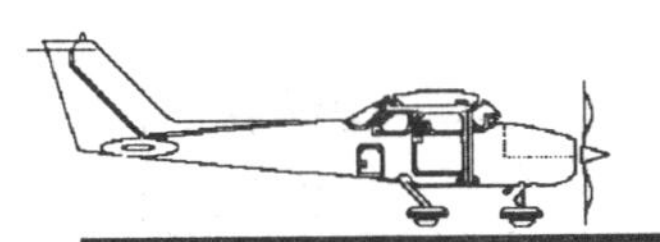
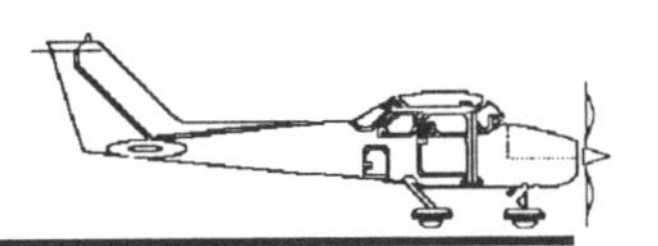

were no injuries to the four people on board.

Nov. 22

Although the engine of a T210F was reported to be sputtering on the takeoff roll, the commercial CFI continued the takeoff and declared an emergency after liftoff. Both fuel caps had deteriorated seals and had leaked water into the filler necks. The fuel strainer and fuel flow divider had evidence of corrosion damage. Water was found in the left reservoir tank to the fuel selector valve. The aircraft had an emergency three days before this takeoff when power was lost after takeoff. Substantial amounts of water were drained after this incident. The aircraft had flown only two hours in the previous eight months. Three days before the accident the aircraft had only one-quarter full fuel tanks, and was topped off. It had not had an annual for two years.

Dec. 6

The pilot of a T210L lost control shortly after takeoff, resulting in an uncontrolled descent. It struck the ground with its left wingtip, causing it to flip over on its side. The pilot reported the autopilot had run away on previous flights, but it was bench tested and found to be normal. The pilot in this Lawton, Miss., accident was 38 and had a private license.

How many of those could have been avoided with better pilot training? Quite a few.

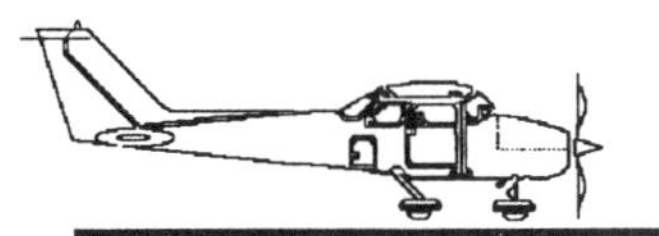
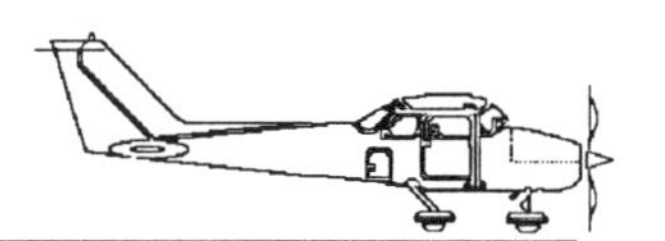

CHAPTER SEVEN

SOME OTHER CESSNA SINGLES

The Hawk XP: Love It Or Leave It

People either love or hate the Hawk XP, says Clark Hanger of Hanger Aviation at East Cooper Airport near Charleston, S.C. He had three XP aircraft for sale at the time of our conversation in late December, 1988. They were all repainted with recent engine overhauls and new interiors.

Hanger said about all one gains from buying an XP, besides five knots airspeed, is a shorter takeoff roll. There isn't much gain in load carrying capability.

"Some people think they are too much fuss," he said, referring to the variable pitch prop and the extra maintenance such a propeller requires.

The airplane has few deep, dark secrets. But be aware the engine on the 1977 model has only a 1,500 hour time between overhaul. Later models have engines that can go for 2,000 hours before overhaul is required. Hanger's all-around favorite single-engine Cessna is the Skylane, he said.

Also, don't lean the engine to get eight gallons per hour of fuel flow, because the higher heat will warp the cylinders, says Charley Tatum, Hanger's business partner in Texas. Tatum falls into the category of one who loves the Hawk XP.

"That is what I would rather be in when flying single-pilot IFR," Tatum said.

Tatum's experience with maintenance costs has been excellent. The most expensive annual inspection he has had cost $625. (The author has experienced several $700, 100-hour inspections with a Cessna 152!) But there is no way around the fact that variable-pitch props cost more to maintain than fixed picth props.

The 172XP Type Certificate

There are no nasty "notes" at the end of the 172XP type certificate to worry about, only one pertaining to airspeed markings and another routine note about weight of unusable fuel.

The airplane was certificated May 28, 1976,

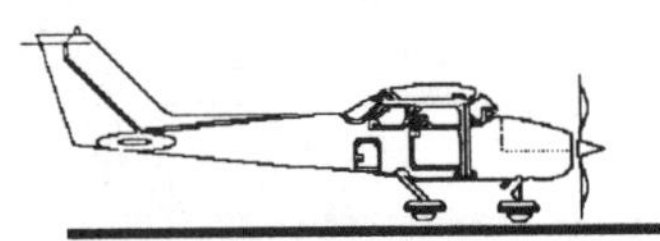
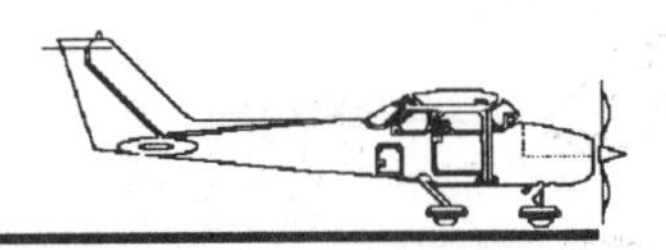

meaning that Cessna offered it beginning in 1977. It uses the Continental IO-360-K or IO-360-KB engine. The grade of gas it burns is 100/130 **for serial numbers R17220000 through R1722724.** It can burn 100LL/100 for serial numbers R1722725 and on.

There are also different maneuvering airspeeds depending on the serial numbers, but we'll get to that in a moment. The point is, not all 172XPs are alike.

The propeller for the landplane version is the McCauley constant speed propeller 2A34C203 hub with 90DCA-14 blades, with a diameter not over 76 inches and not under 74.5 inches. The floatplane uses the McCauley constant speed 2A34C203 with 90DCA-10 blades. The diameter is not over 80 inches and not under 78.5 inches.

As mentioned, airspeeds are slightly different depending on what serial numbers you are talking about. For serial numbers 680, R1722000 through R1723199, maneuvering speed is 105 knots. But for serial number R1723200 and on the maneuvering speed is 104 knots. Other speeds are the same for all serial numbers: max structural cruising speed is 129 knots, never exceed is 163 and flaps extended is 85.

So what would you expect to see on an actual trip? A speed of 122-125 seems reasonable.

The ramp weight of the plane can be 2,558 pounds for serial number R1722930 and on. The max weight for normal category, landplane and floatplane, is 2,550 pounds, but utility category is only 2,200 pounds. So obviously most of us would want the normal category Hawk XP.

Max baggage is 200 pounds. Fuel capacity is 52 gallons, with 49 usable. It uses eight quarts of oil, with five quarts usable.

Analysis of Hawk XP Accidents

An interesting thing happened in the search for 172XP accidents with mechanical causes: the National Transportation Safety Board, in a computer search for this book, couldn't find any landplane incidents! Only one, a failed float on a floatplane, could be classified as mechanical.

"But there must be some!" a reporter protested. "Look harder."

So, the very helpful lady handling the request searched 1983, 1984, 1985, 1986 and 1987. She came up with a very thin pile of computer printout sheets. None of them was due to a mechanical cause!

In **January 1983** a pilot with 68 total hours tried to land with the wind in Harmony, Wyoming. It didn't work and he hit a fence.

In **July, 1983,** a pilot in New York pushed the throttle by mistake after startup, and hit a parked airplane.

In **August 1985** at International Falls, Minn., a float failed during takeoff. The float assembly was disconnected. The aircraft had substantial damage, but the pilot was not injured as the airplane collided with the water.

In **February 1986** a pilot with a blood alcohol content of .175% at Napa, Calif., after difficulty with taxi and takeoff instructions, finally took off, tried to return and hit a tree. The pilot was killed and his passenger was unconscious for two months.

In **April 1987**, at Superior, Wis., a pilot landing in a 50-degree, 12-knot crosswind was blown off the runway and nosed over in mud. Neither the pilot nor the passenger was hurt.

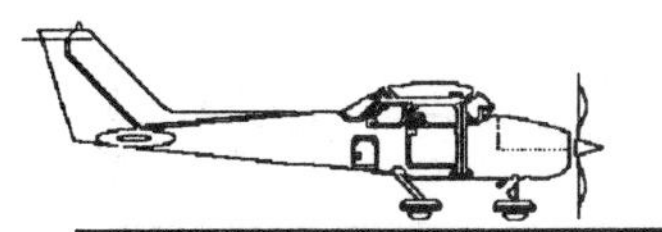
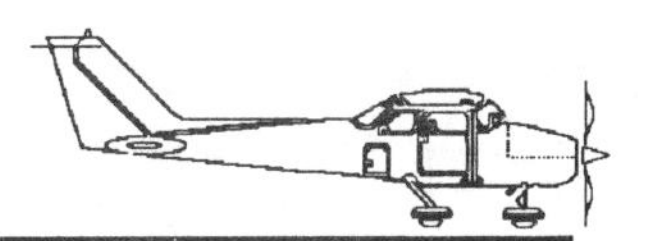

Some Hawk XP Maintenance Costs

Let's look at some actual bills for a Hawk XP based at Dulles International Airport near Washington, D.C.

In February the aircraft, a club plane, needed a jump start for $20, followed by a new battery which cost $177.65. The 100-hour inspection, required because the plane is in commercial service, was only $233.94. It was done in nearby West Virginia, where costs are a little cheaper.

IFR certification cost $132.86. There was also a $16 expense for door stops. Tiedown, just for interest's sake, was $173 a month at Dulles.

In March an alternator needed replacement, and that cost $438.71. The cylinders needed "work," which was $408. A rough month, maintenance-wise. Another 100-hour inspection cost $427.51. By the way, the log books show frequent clogging of the oil screen throughout the year. Something to keep in mind about the Hawk XP.

In June the Automatic Direction Finder needed repair, costing $134.72. An annual inspection and a new starter, push rods, seals and gaskets came to a huge $773.

Brake linings were a little better, at $27.80. A tire was $93.60, while a vacuum pump cost only $33 to repair.

Commercial insurance was between $3,000 and $4,000 a year, but a private owner using the plane for his own satisfaction, with 1,000 hours would pay only $700 to $800 per year.

Hawk XP Airworthiness Directives

Nothing shocking resides in the airworthiness directives for the Hawk XP, also known as the R172K. Biggest problems owners mentioned are: (1) the engine burns NOT LESS THAN 10 gallons-per-hour, and (2) the oil screen in the engine is forever getting blocked. (Owners who run their engines leaner than 10 gallons per hour risk damage from high temperature, says a Hawk XP operator in San Antonio.)

A check with the Federal Aviation Administration District Office at Dulles International Airport, Va., shows no airworthiness directives that affect the Hawk XP alone. All of those found affected several Cessna models. In 1987 Cessna warned, for example, that the seat locking mechanism on most of their single-engine piston aircraft was not locking.

In 1986 a directive was issued to prevent **engine power loss due to loss of attachment of the engine controls**. Aircraft affected included 150s, 152s, 172s, 185s, 206 models, and 210s.

Also in 1986 Cessna warned Hawk XP owners, and owners of most other piston Cessnas, to place a container under the fuel drain at the engine and **drain for at least four seconds to check for water**. The same was recommended for the wing drains. (Some models of aircraft, particularly Pipers, require six seconds for water in the fuel lines to reach the engine during draining.)

In 1983 an airworthiness directive warned about **possible loss of aileron hinge pins** because cotter pin holes are not correctly located.

Most models of 172s, including the Hawk XP, were given an AD in March 1983 to prevent **possible jamming of the elevator control.** It required modification of the right-hand control wheel yoke.

In September 1981 there was another AD about the elevator, this one to ensure **integrity of**

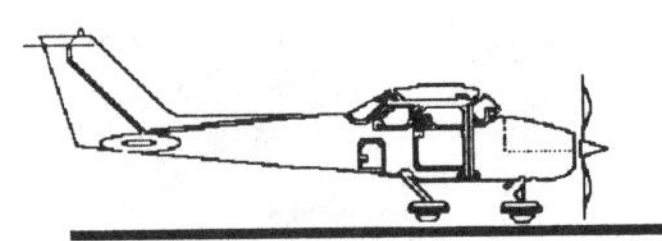
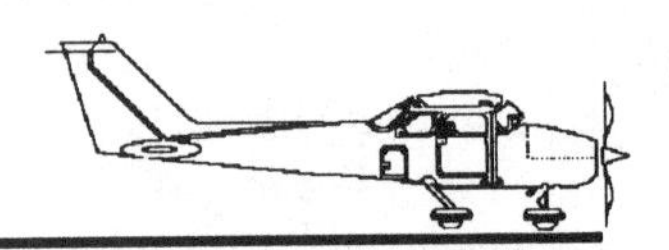

the elevator control system which required cleaning the mating surfaces of the bellcrank and the elevator cable clevises (forks).

In January 1981 there was an AD for 152s and 172s, including the XP, to reduce the possibility of **fuel depletion due to incorrect fuel quantity markings.** The answer seemed to be verification of fuel gauge readings with visual inspections, but pilots are supposed to do that anyway during preflight.

A 1980 AD looks a little more serious than any so far. It was to preclude **failure of the engine oil pressure pump drive shaft, and resulting oil pressure loss** caused by an improperly installed tachometer shaft connector. It required removal of the Cessna tachometer drive adapter assembly from the engine accessory case.

Also in 1980 there was an AD to assure continued **structural integrity of the wing flap direct cable,** "thereby preventing possible sudden unexpected retraction of the left wing flap." It affected the 150, 152 and 172 including the XP.

Nearly every 172 model year, and the XP, was affected by an August 1980 AD to preclude the possibility of a **fuel leak or an in-flight fire due to contact between a map light switch and an adjacent fuel line.**

Speaking of fire, a 1979 AD warned of action needed to prevent an **in-flight electrical system failure, smoke in the cockpit and a fire in the wire bundle behind the instrument panel.** The solution was to place a circuit protection device in the wiring to the cigar lighter.

In 1977, there was an August AD to prevent **unwanted propeller rotation when the external electrical ground power receptacle is used.** Sometimes when power is hooked up, but the starter is not engaged, the propeller starts turning.

The Cutlass RG: A Trainer or a Real Plane?

There are those who feel the Cutlass RG was intended mainly as a trainer, and should stay that way. A dealer in Charleston, S.C., said most of his sales of Cutlass RGs are to flight schools which want a complex aircraft that is easy to operate. The gain in speed is only about 10 knots (others claim 5 knots in actual experience), but the cost of maintenance goes up disproportionately due to the variable pitch prop and retractable landing gear.

There is another Cutlass that makes a good cross country plane, but you probably won't find one. It is the "Cutlass," period. No "RG" after the name. It is fixed gear and fixed prop. Most pilots do not even know there is such an animal. It is basically a 172 with a 180 hp engine; goes like crazy.

The disappointing thing about the Cutlass (fixed gear version) is its scarcity. Judy Wood of the Cessna Owner's Association (1-800-247-8360) says there are only 29 registered in the U.S.! Judy is helpful on the phone to prospective buyers, and her organization is a good one to join before and after buying.

As this was written the Federal Aviation Administration issued a note of concern about the doorposts of the Cutlass which hold the windshield. The doorposts on all 172s after 1980 were thinner. In late 1988 a windscreen blew out of a Cutlass, leaving the pilot to face the breezes. (He gained better control by opening the doors to let the wind out of the cockpit.)

The 172RG Type Certificate

Fortunately, there are no serious notes attached to the 172RG type certificate, which was approved June 1, 1979 in time for the 1980 model year. The two notes mentioned refer to routine center-of-gravity calculations and airspeed markings.

If there are any variances from the type certificate, then a supplemental type certificate must be on file with the owner.

The engine is the Lycoming 0-360-F1A6. It uses 100 LL and 100 minimum grade gasoline. The propeller is the McCauley constant speed B2D34C220 hub with 80VHA-3.5 blades. The diameter is not over 76.5 inches and not under 75.5 inches. The governor is the McCauley C290D3/T18.

Maximum structural cruising speed is 145 knots, but the landing gear can be extended up to 164 knots. Max weight is 2,650 pounds, but it can be 2,658 on the ramp if you burn off the fuel before takeoff. Max baggage is 200 pounds.

Oil capacity is eight quarts, while fuel is 66 gallons, with 62 usable. It was manufactured from 1980 to 1985.

Wing flaps extend 30 degrees. Ailerons travel up 20 degrees, and down 15 degrees. The elevator goes up 28 degrees, and down 23 degrees. The rudder goes 16 degrees to either side.

Accident Reports for the 172RG

The National Transportation Safety Board did not find a great many Cessna 172RG accidents when asked to check both 1987 and 1986. Many they found were not the fault of the plane.

May 22, 1987. For example, on May 22, 1987 at Ozark, Ala., a 172RG crashed in the trees near runway 12 at Blackwell Airport after running out of fuel. An instructor was checking out a pilot in the aircraft.

NTSB said the student did not understand the operation of the fuel selector, apparently not realizing the handle could not be stopped outside the detent position. The NTSB records are not entirely clear on that point. The instructor failed to catch the problem. The selector had been moved from "right" to "both" during the pre-landing check.

May 25, 1986. The NTSB classified as "non-mechanical" an accident on May 25, 1986 at Bridgeport, N.J., in which a 30-year-old single- and multi-engine qualified pilot continued a take-off roll as the 172RG emitted black smoke and backfired.

The backfiring continued as the aircraft became airborne. It could not climb or even maintain flying speed. It crashed on a public roadway and caught fire. All four persons on board suffered serious injuries and the aircraft was destroyed. The NTSB said the pilot, who had 715 total hours and an instrument rating, was "over-confident."

May 23, 1986. A 35-year-old woman flying at Hawkins Field, Jackson, Miss., landed gear up after discovering the nose wheel doors were jammed and would not allow the nose gear to lock down. Examination of the nose landing gear revealed the gear-up stop bumper fell off, allowing the nose gear to retract into the well too far and jam in the gear up position. Pretty good thinking for a pilot with only 181 hours total. She was 29 hours into her instrument rating.

Another similar incident has been reported on

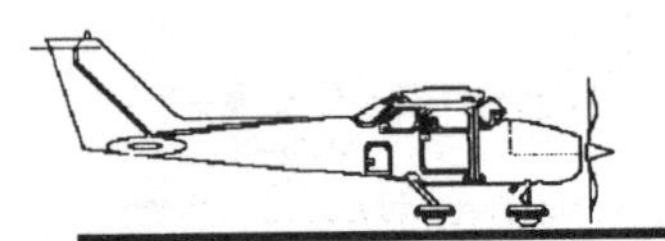

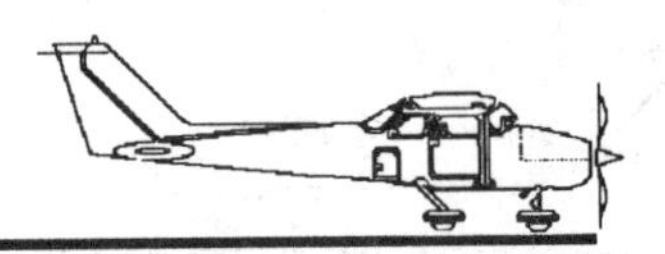

another Cessna model in this book.

April 7, 1986. A St. Paul, Minn., pilot, aged 51 with 1,148 hours, discovered smoke entering the cabin after passing through 4,000 feet. As he declared an emergency and turned toward St. Paul Downtown Airport, the engine began to shake violently.

Although the pilot was cleared to land, he continued to a point four miles past base leg and attempted to fly under a high-power line. He struck it, and elected to continue with a gear down landing in soft soil. During rollout, the nose gear collapsed. He had 370 hours in a 172RG, 246 hours instrument time and 199 multi time. He was also a flight instructor.

The NTSB found that the oil regulator was siphoning due to failure of an oil seal. All the oil was lost. NTSB further found a maintenance inspection of the aircraft was improper.

Cessna 172RG Airworthiness Directives

The 172RG has the standard airworthiness directives shared by several Cessna models, like seat locks and shoulder harness belts that slip. But there appears to be nothing outstanding, like the 172's "H" model engine, that raises a red flag. In fact, the 172RG has the Lycoming 360 engine, considered by many to be the best engine flying.

Here is a partial listing of airworthiness directives for the 172RG. The recurring ones are the ADs that keep you paying maintenance bills on into the future.

In 1983, for example, there was a recurring AD on the possibility of carbon monoxide contamination entering the cabin. In 1984 there was a recurring AD on induction air filters, to prevent possible engine power loss or stoppage caused by engine ingestion of filters.

In 1980 there was a recurring AD on Stewart-Warner oil coolers.

Looking over the more serious ADs the 172RG has had, but which are not recurring, we see:

1981 - possible failure of the aluminum air filter retainer screen or gaskets;

1984 - procedures to reduce valve sticking in Lycoming engines;

1984 - nose gear actuator spring;

1982 - warning against use of automotive gasoline (this one keeps popping up, and people keep ignoring it);

1982 - propeller blade shank inspection;

1982 - landing gear power pack coupling improvement;

1983 - landing gear power pack grounding improvement;

1983 - main gear downstop pad improvement;

1981 - Slick magneto coil inspection; and,

1980 - nose gear actuator clearance.

Some other ADs include a 1983 warning that failure to secure the fuel cap securely, or to use fuel caps with improper sealing, can cause loss of fuel and erroneously high fuel quantity indications.

In 1983 an AD warned of possible loss of

aileron hinge pins, but it applied to many models of Cessnas.

In 1981 AD 81-14-06 was issued to ensure the integrity of the rudder trim/nose gear steering bungee. Also in 1981 the AD numbered 81-05-01 warned of the possibility of fuel depletion due to incorrect fuel quantity markings.

In 1980 an AD numbered 80-01-06 was issued to preclude overtravel of the flap actuator assembly and subsequent flap system failure which can result in an asymmetric flap configuration.

The 1982 AD issued on McCauley propellers and affecting several Cessna models was issued to prevent possible propeller bland shank failure. The answer was to require dye penetrant inspections to find cracks or forging "folds" in the metal. Paint and blade annodizer had to be removed for the inspection. This is AD 82-27-02.

The answer to AD 81-05-05, concerning carbon monoxide entering the cabin, was to visually inspect the cabin heater shroud on the engine exhaust muffler to assure it had not slipped off its flanges. The shroud had to be removed and trimmed as well.

Finally, AD 80-19-08 warned that the mixture control wire could slip, resulting in power loss due to a lean mixture. The answer to that problem included visually inspecting the right side of the carburetor to see that the mixture control connection is proper.

Cardinal, Skykwagon/Stationair

Included here is abbreviated information on the Cardinal and Skywagon/Stationair without giving them the full treatment. The reasons are; (1) dealers raised some concerns about the Cardinal resale value, and (2) few people need the Skywagon/Stationair. If that changes, future editions of this book will look at them in more detail.

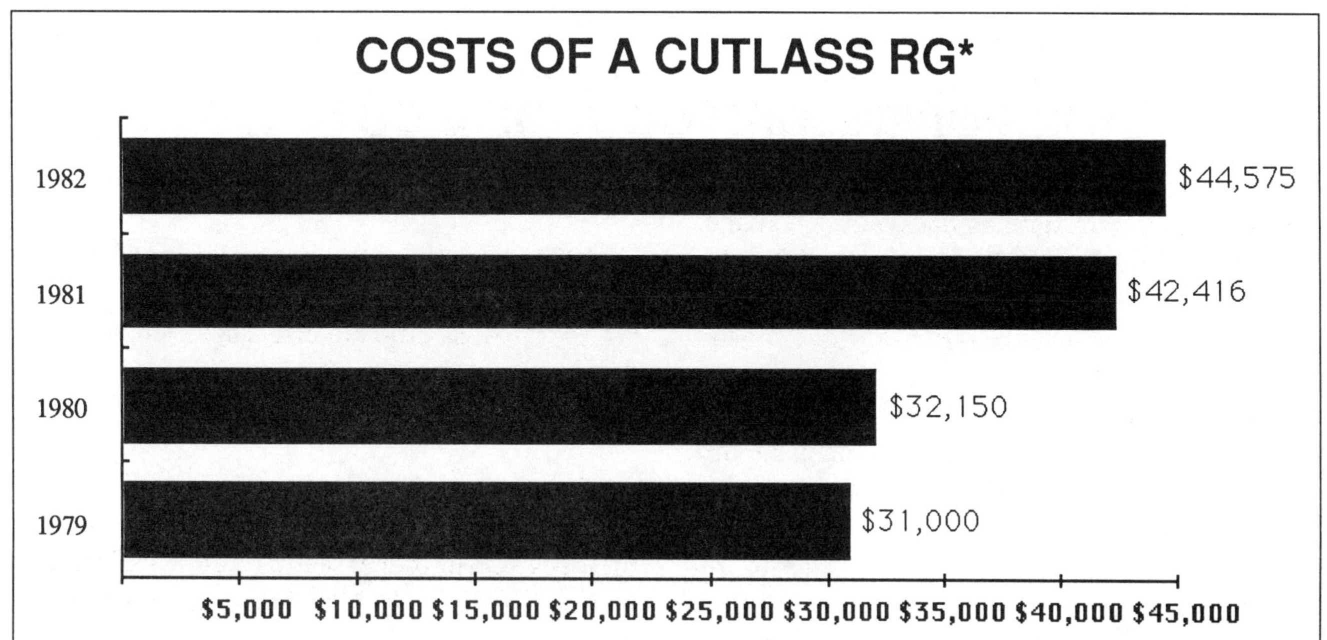

* Matt Hagans, owner of Indiana Aircraft Sales, says the Cutlass RG is not as fast as a Piper Arrow, but the engine is a "good performer."

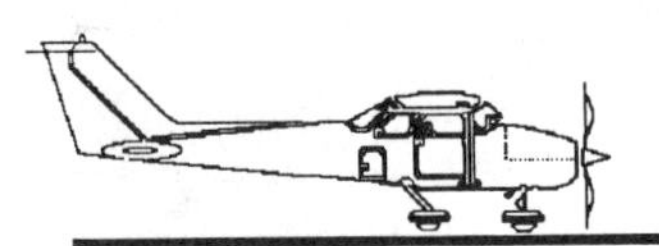

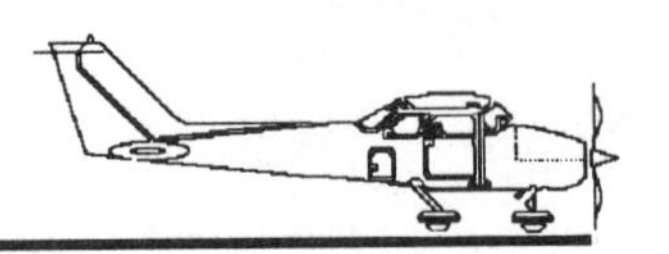

The Cardinal is a beautiful plane to look at, and cheap to buy. But if it is cheap to buy it also doesn't bring much when it is time to sell. The reason, officials at Indiana Aircraft Sales in Indianapolis said, is that Cessna discontinued the Cardinal prior to the general cutback in airplane production. Resale value on a discontinued aircraft will always be lower than for aircraft that were in production when the general aviation airplane market brought Cessna piston manufacturing to a halt.

So the Cardinal is not a good investment, solely because of resale value. In addition, there is general agreement among those interviewed that the 150 hp. Cardinal was an under-powered mistake. Finally, there is concern that the buyer who gets a Cardinal RG because it is only $28,000 or so, cannot afford maintenance on a variable-pitch prop and retractable gear.

One source indicated the doors of a Cardinal often seem to fit poorly. The reason? It has a weaker airframe due to greater use of molded plastic, so the door frame is often misshapen.

But the airplane still looks great, and that goes for the split-level instrument panel inside, too.

Skywagon/Stationair Loves Gas

The Skywagon/Stationair had those problematic bladder fuel tanks until 1979—the tanks with the wrinkles that capture water. Those tanks were also a problem in the 210 and 182. After 1979, of course, Cessna went to a wet wing, eliminating the bladders. The Stationair, by the way, carries up to eight people.

The bigger concern is the 15.8 gallons of gas the plane consumes every hour. Sure, the 210 also has a sweet tooth for gas, but at least you get plenty of speed in return. Operating costs, one dealer said, include: 15.8 gallons of gas per hour at 76% power ($26.86); $1.03 an hour for oil; $11.52 for maintenance; $8.61 as a reserve for engine and prop overhaul; and the usual $5 an hour for avionics (assuming there is $5,000 worth of avionics in the plane). That comes to $53.02 an hour to move yourself and five (or seven) other people 150 miles. Not bad, considering it is more

PER HOUR OPERATING COST OF A CUTLASS RG*

Fuel — 10.2 gal/hr at $1.69/gal.	$17.24
Oil	.87
Maintenance	12.16/hr
Engine/propeller overhaul fund	5.50
Avionics .1¢ per hour times cost	6.00
TOTAL COST	**$44.77**

* Information from Matt Hagans of Indiana Aircraft Sales, Indianapolis, Ind.

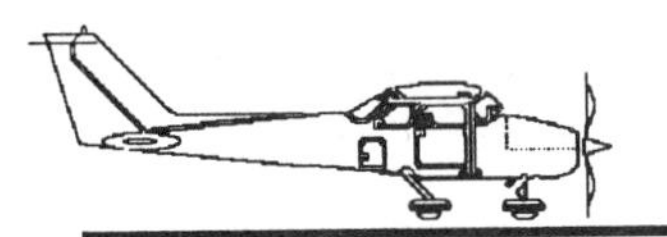
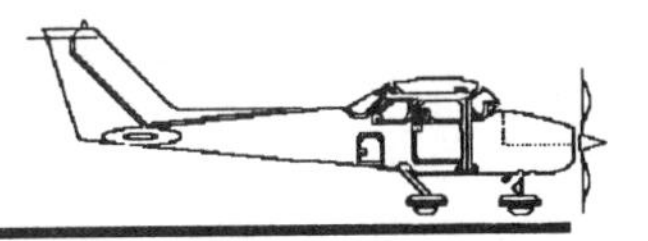

comfortable than a car and three times as fast.

A Look at Insurance

Basically, the Skywagon/Stationair and Cardinal RG are about $1,000 more per year to insure than a Cessna 172. The straight-leg Cardinal is about $1,400, or double what Cessna 172 insurance costs. To get these figures, Avemco was called and given an example of a 1,000-hr. pilot with no time in either the Stationair or Cardinal/Cardinal RG, but 100 hours of retractable time and an instrument rating.

A word about insurance company requirements is in order. Insurance companies don't give you a lower rate for having "about" 100 hours of retractable time. You must have exactly 100 or more. If you fall two hours short, in other words, go out and get the two hours in a rental aircraft. It means big savings.

There may also be reductions for 50 hours of retractable time. You get further reductions if you have 250 or 500 hours of retractable time, of course. So ask the insurance company for experience levels required to reduce the rates. But meet the requirements exactly.

177/177RG Cardinal Type Certificate

First, the good news. There are no notes for the straight leg Cessna 177 at the end of the type certificate that are going to frighten anyone. The notes concern placards that must be in the airplane, tell which models have a 14-volt electrical system and which have a 28-volt system, and provide the location of cylinder head probes.

The type certificate was first approved Feb. 16, 1967, for the 1968 model year.

The bad news is that this Cardinal has only 150 hp. That turned out to be too little power for it. The engine is the 150 hp. Lycoming 0-320-E2D, which takes 80/87 minimum grade gas.

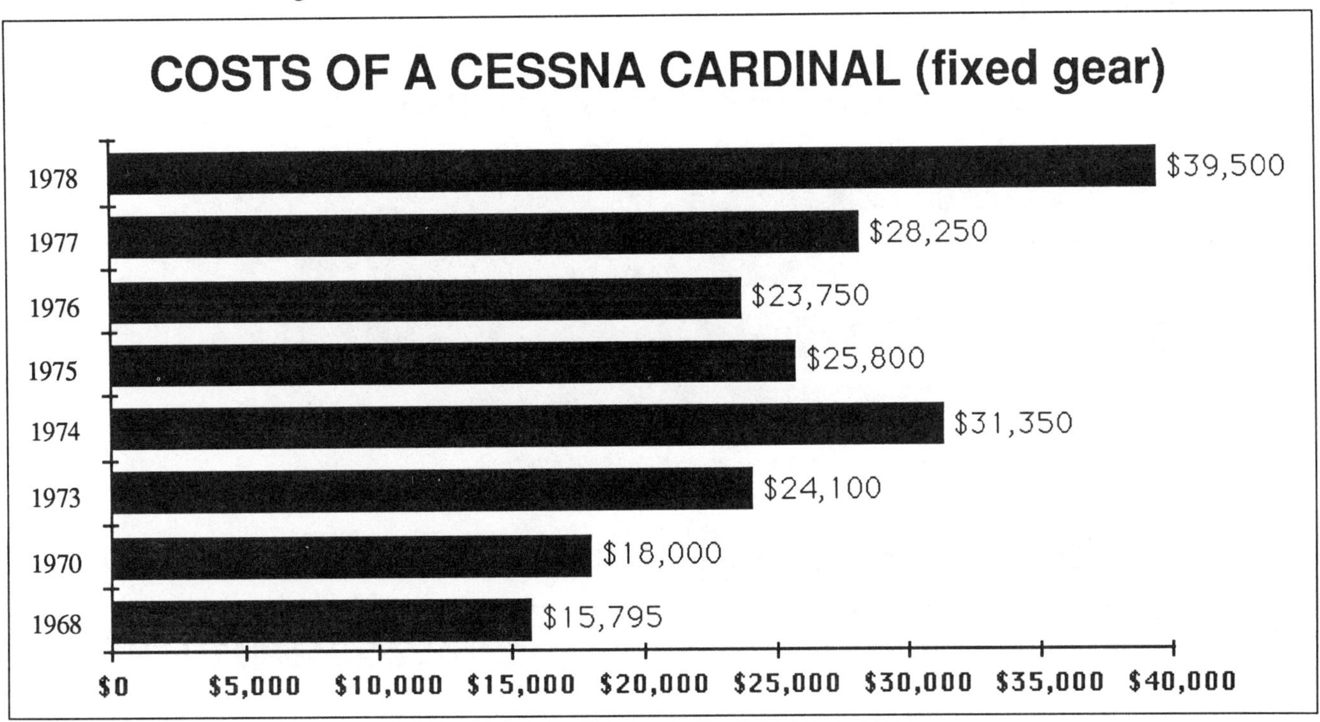

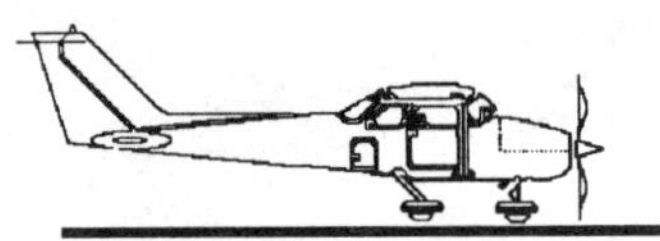

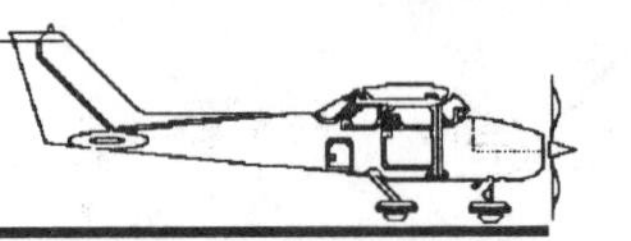

Max structural cruising speed is 145 mph, while flap extension speed is 105 mph. Max weight for the four-seater is 2,350 lbs. Max baggage is 120 lbs. The propeller is the McCauley 1C172/TM, which can be not over 76 in. and not under 74 in.

Fuel capacity is 49 gallons, with 48 usable. Oil capacity is eight quarts.

On June 28, 1968 a new type certificate was approved for the 177A, this one with a bigger engine, the reliable 180 hp. Lycoming 0-360-A2F, which takes 100/130 grade gas.

The propeller became the McCauley 1A170/EFA. Now, with the larger engine, the structural cruising speed limit became 150 mph. Flap extension stayed the same, at 105 mph., as did the never exceed speed of 185 mph.

The max weight increased to 2,500 lbs., but baggage remained at 120 lbs. for balance considerations. Fuel capacity remained the same at 49 gallons, with 48 usable.

This model and the first one above have fixed-pitch props.

The first variable pitch prop came with the type certificate for the Cessna 177B, granted July 28, 1969 in time for the 1970 model year.

The airplane still had a 180 hp. engine, but this time it could also take 100LL gas. The engine is the Lycoming 0-360-A1F6 or 0-360-A1F6D.

The prop is the McCauley 2D34C202/82PA-6, or the McCauley B2D34C208/82PA-6, or the B2034C211/82 PCA-6.

Structural cruising speed with the new prop was now a maximum of 155 mph, only 5 mph. above the fixed-pitch prop speed. Flap extension remained at 105 mph.

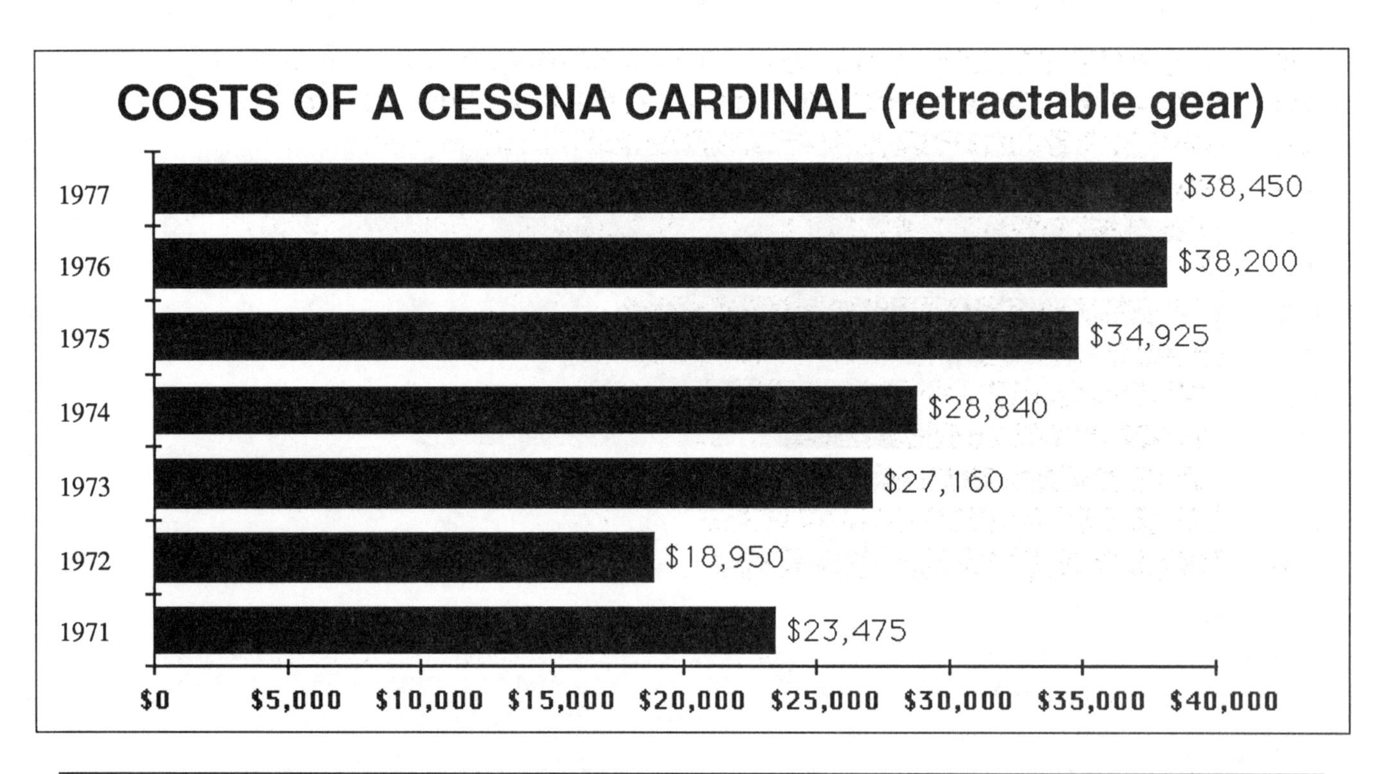

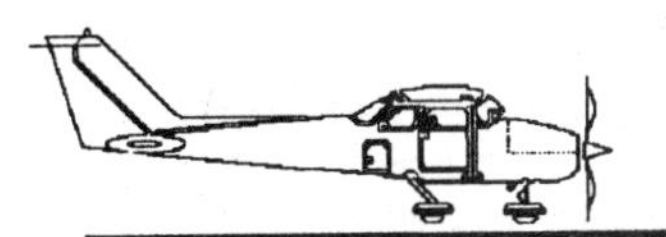
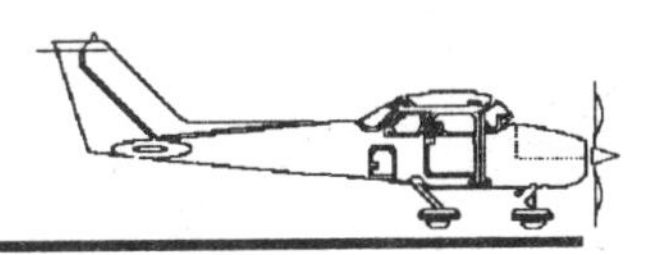

Fuel capacity went up to 50 gallons, 49 usable, but the max weight remained at 2,500 lb.

The Cardinal RG Type Certificate

The first Cardinal RG type certificate, for the 177RG, was approved Aug. 11, 1970 in time for the 1971 model year.

This time, the engine is the Lycoming IO-360-A1B6 or IO-360-A1B6D, rated at 200 hp. The propeller is the B2D34C206/78TA by McCauley, or the B2D34C207/78TCA.

Max structural cruising speed is up another 5 mph., this time to 160 mph. Landing gear operating speed is 140 mph. and flaps extended speed is up 5 mph. to 110 mph. However, on serial numbers 177RG0419, 177RG0788 and up, the max structural cruising speed jumps to 142 knots (it is 139 knots above). Landing gear operating speed is 125 knots (it is 122 knots above).

Max weight is now 2,800 lb., but baggage remains at 120 lbs. Fuel capacity is 51 gallons, with 50 usable. But now, larger tanks are also available with 61 gallons, 60 gallons usable.

The weight and fuel limits apply to all RGs, regardless of the serial numbers mentioned above.

Again, the notes at the end of this type certificate contain nothing serious. They make reference to numerous placards required to be in the airplane — so many in fact that the plane that is properly placarded must look like a billboard.

Cessna 177 Accidents, Fixed Gear and RG

As compared to the 172RG, the 177 model had lots of accidents in 1986 and early 1987, the most recent period for which NTSB had com-

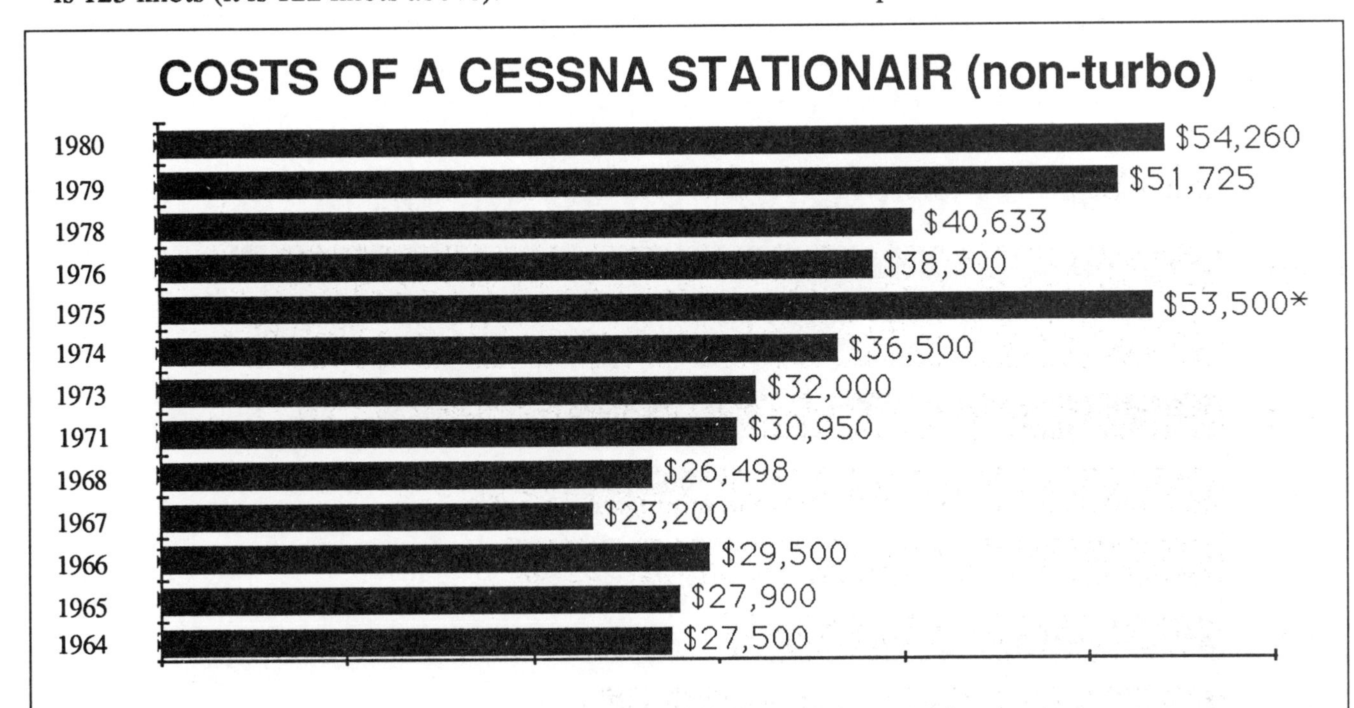

* Information from a recent issue of *Trade-A-Plane*. The 1975 model is either overpriced, newly restored or very well equipped.

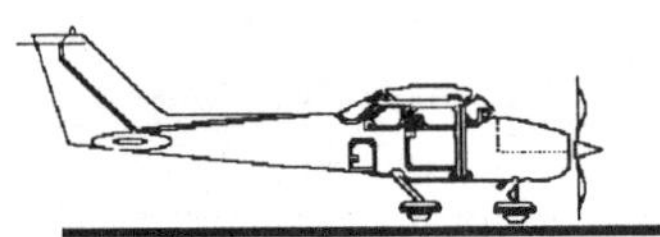
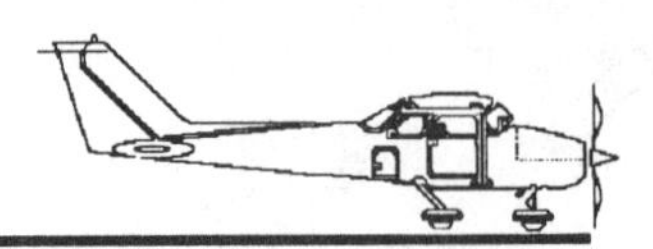

pleted full records (as of December 1988). Here's the list:

April 18, 1987. Brookville, Ohio. A 55-year-old pilot with 6,386 hours, including 1,770 in multi-engine aircraft, suffered a power loss during climbout at an altitude of 30 feet. He was flying a 177 fixed-gear aircraft with the 150 hp. engine.

He elected to land straight ahead, like the book says, in a soft, plowed field. The aircraft nosed over on landing with substantial damage. There were no injuries. The NTSB found water in the fuel and said the pilot had done an inadequate preflight.

Sept. 15, 1986. Westminster, Md. The 63-year-old pilot experienced the loss of five inches of one propeller blade tip while cruising at 3,500 feet. The aircraft went into severe vibration and partial power loss. The pilot set up an approach at Clearview Airport, which has a 1,830 foot, hilly runway. He touched down midway, ran off the end and nosed over in a ditch. He was flying a fixed gear, 180 hp. 177B with the Lycoming 0-360 engine. His total time was 473 hours. NTSB was unable to determine why the propeller failed, but it listed the pilot as misjudging his touchdown point.

Sept. 6, 1986. Springfield, Va. A 36-year-old pilot flying in conditions conducive to carb ice experienced roughness in the engine, then it smoothed but later quit. He landed in a parking lot, killing one driver and receiving slight injuries. The NTSB found no carb heat was used during the emergency. It was a fixed-gear 177B, with the Lycoming 0-360 engine (of 180 hp.).

Aug. 1, 1986. Oshkosh, Wis. A 58-year-old pilot with 784 hours put his airplane in Lake Winnebago after running out of fuel on his way to the Oshkosh Air Show. It was a 177RG, and there were no injuries.

July 20, 1986. Dania, Fla. A 35-year-old pilot with 165 hours, flying a 177RG that had received an overhauled engine just 1.1 hours earlier, lost power while flying along the beach.

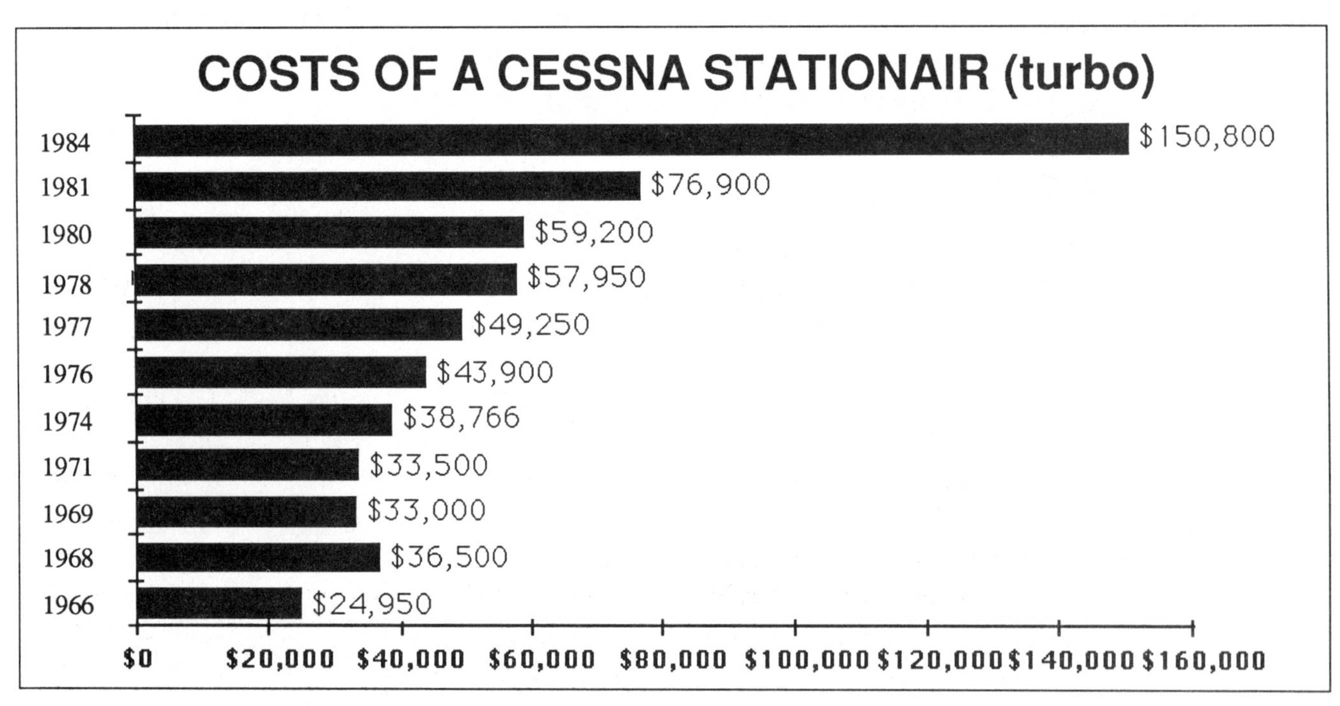

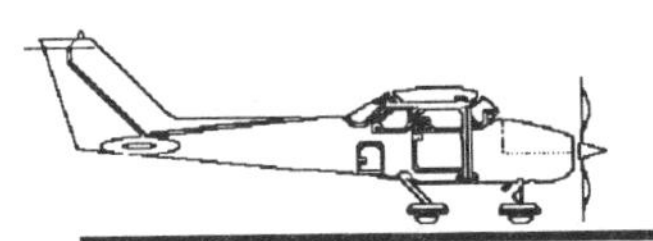
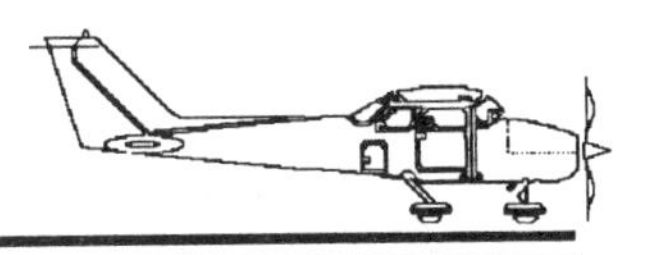

All four aboard had minor injuries. The fuel line from the fuel servo unit to the fuel divider block was loose. There was evidence of fuel spray patterns on the inside of the lower engine cowling. An annual inspection was done at the same time as the engine installation.

July 3, 1986. Huntsville, Ala. The pilot of the 180 hp., fixed-gear 177A took fuel samples before takeoff at Moontown, but found no water. But when the 61-year-old pilot landed in a soy bean field with an engine failure, NTSB found the carburetor was full of water. The aircraft had been left on the ramp with tanks empty for one week, and there was an extreme temperature variation the night before the flight. After taking the fuel samples, the 1,182-hour pilot added 20 gallons of fuel to each tank but did not take additional samples. The fuel supply was not contaminated. The engine quit at 500 feet.

June 13, 1986. Odessa, Texas. The pilot had been drinking before the flight (need we go on?) but took off anyway. He was 40 and had a total of 190 hours, including 100 in his fixed-gear, 180 hp. 177B. He decided AFTER takeoff he was not in any condition to fly, and returned to the airport. His seat belt, which had gotten caught in the door before takeoff, caused the door to open during the return. This distraction caused him to touch down short of the runway at Odessa-Schlemeyer Airport,. He then skidded 255 feet. It turned out the seat belt wasn't even fastened. He was thrown out and seriously injured. His blood/alcohol level was .29%.

May 30, 1986. Jacksonville, Fla. The 61-year-old pilot, with 3,957 hours and 150 in the 177RG, insisted to investigators after the accident that the gear was down and locked. But the gear collapsed and the plane slid to a stop at Craig Municipal. After the crash, the gear operated normally in all modes. But the gear warning horn was inop. The NTSB said the pilot failed to follow the checklist.

April 29, 1986. Placerville, Calif. A 38-year-old pilot was flying the 177RG one day after a 100-hour inspection. While enroute to San Francisco, the engine began to overspeed and the propeller control was not effective. Then oil pressure went to zero and the pilot landed in the Ice House Reservoir. The governor oil pressure line was chaffing against the engine crankcase and broke. It was not secured with clamps as required by the Avco Lycoming maintenance manual. It was the Lycoming IO-360 engine generating 200 hp., the NTSB report says.

April 12, 1986. Dallas, Texas. A 34-year-old pilot with 2,200 hours, who had a friend waiting at Addison Airport in north Dallas, saw severe weather over the airport. He was advised three times by a controller to divert to Love Field in downtown Dallas. He flew into the storm, and lost the windshield to a hail storm. He may have been incapacitated by the hail, but Cessna engineers said the plane is uncontrollable without the windshield. It was a 177RG. The pilot, alone in the plane, was killed.

April 6, 1986. Orlando, Fla. The 48-year-old pilot said the gear on his 177RG extended during approach to Orlando Executive, but collapsed at touchdown. The alternator belt was loose, allowing the battery to drain enough so that the hydraulic pump could not fully extend the landing gear.

March 29, 1986. Aztec, N.M. The engine of a fixed-gear 177B, a 180 hp. Lycoming 0-360, failed after the rubber gasket surrounding the air filter deteriorated and lodged in the carburetor venturi. The 38-year-old pilot had 1,509 hours, including 1,090 in make and model. He landed on a dirt road, hit a ditch and overturned. Examination revealed neither AD 81-15-03 nor super-

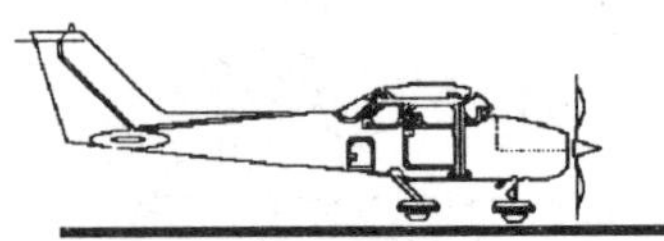
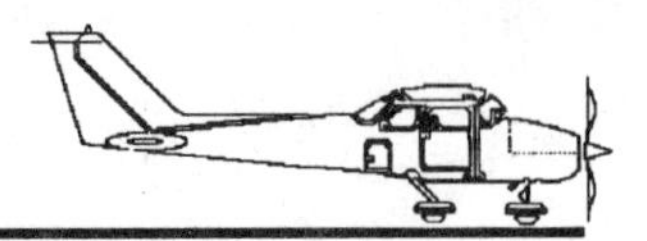

ceded AD 78-25-05 had been accomplished.

March 6, 1986. Anaconda, Mont. Examiners of this 177 fixed-gear crash found the carburetor bowl clean, dry and free of any fuel. The pilot and two passengers were killed, but the seriously injured passenger said the plane lost power during cruise over mountainous terrain. The pilot hit a tree during the forced landing. Approximately five gallons of automotive fuel were drained from the wing tank. A fuel sample from the pilot's storage tank at Thompson Falls, Mont., showed water contamination. The aircraft had been at 9,000 feet, in below freezing temperatures, for just over an hour. Engine disassembly showed no evidence of failure.

Cardinal Airworthiness Directives

These are a few of the airworthiness directives that seem most worrisome to the present or potential Cardinal owner. ADs that apply to most Cessna models are not included, such as the well-known seat locking mechanism failure and slippage of the shoulder harness.

December, 1977. To prevent unwanted propeller rotation (when using external ground power receptacle). All this AD required was to apply power to the external ground power receptacle and see if the prop turns. If it doesn't, you have complied with the AD.

June, 1976. To prevent loss of engine oil through the engine oil cooler fluid fitting side.

March, 1976. To preclude restrictions of control movement due to jamming of the ARC PA-500A actuator gear train.

February, 1975. To preclude separation of the foam rubber air filter seal.

June, 1974. To prevent failure or chafing of the oil pressure gage line located between the firewall and the oil pressure gage, install clamps between the flap control cable and the oil pressure gage line in accordance with Cessna Service Letter SE74-2 dated Jan. 25, 1974.

February, 1968. To prevent failure of the oil pressure gage line between the engine crankcase and aircraft firewall.

March, 1971. To prevent cracks in the stabilator attachment angles P/N 1712108.

April, 1971. To detect leakage of flammable fluids from flexible hose assemblies in the engine compartment.

March, 1972. To prevent inadvertent retraction of wing flaps and to insure positive operation of the electrical wing flap actuators.

September, 1968. To prevent oscillation in the longitudinal control system [modify the flap system].

Recurring airworthiness directives, those that came out a long time ago but still require re-compliance every so often, include:

➤the 1987 seat locking mechanism AD,

➤the 1986 contaminated fuel AD,

➤the 1982 Bendix distributor gear electrode AD,

➤the 1980 Stewart Warner oil cooler AD,

➤the 1978 Bendix magneto impulse coupling failure AD,

➤the 1977 Hartzell bland shank cracks and blade failure AD,

➤the 1976 Bendix ignition switch AD,

➤the 1976 jamming of the actuator gear AD,

➤the 1974 Leigh Sharc 7 emergency locator transmitter lithium battery AD,

➤the 1972 inadvertent wing flap retraction AD, and,

➤the 1968 stall warning system AD.

So you see, when you buy an airplane that advertises "All ADs complied with," it ain't necessarily so. It should read, "All ADs complied with up to time of purchase." You have to comply over and over at specified time periods.

The 206 and 207 Type Certificates

There are no nasty notes at the end of the certificates for the 206 model. They concern the placards that must be on the airplane and discuss weights and electrical system information.

The 206 certificate was first approved July 19, 1963 for the following model year, 1964. It was called the Super Skywagon.

The engine is the Continental IO-520-A, rated at 285 hp. The propeller is the McCauley D2A34C58/90AT-8 for landplanes and D2A34C58/90AT-2 for floatplanes.

Maximum structural cruising speed is 170 mph., and max weight is 3,300 lbs. for land and 3,500 for a floatplane.

It has six seats and carries 65 gallons of gas, 63.4 usable. It requires 12 quarts of oil.

The U206 was certificated Oct. 8, 1964, and has the same engine as above. Speeds and weights are the same. Fuel and oil capacity are also the same.

The U206A was certificated Sept. 24, 1965. It has the same engine as above, but can use the McCauley D3A32C79/82NK-2 prop or the D3A32C90/82NC-2 prop. Airspeeds are the same, but weights are: 3,600 lbs. for a landplane, 3,500 lbs. for a floatplane or 3,300 lbs. for a skiplane. Fuel and oil are about the same, but this model has 63 gallons usable. Cessna took away .4 gallon.

The P206A was certificated Sept. 24, 1965, while the P206B was certificated Aug. 3, 1966. They have the same engine as above. Now, however, there are numerous propellers to choose from — all made by McCauley. The max cruising speed, max weight, fuel and oil are as above.

The TU206A was certificated Dec. 20, 1965, while the TU206B was certificated Aug. 3, 1966. Now the engine is the Continental TSIO-520-C, with 285 hp. Max cruising speed is still 170 mph., and max weights are also the same.

The TP206A was approved Dec. 20, 1965, while the TP206B was approved Aug. 3, 1966. It has the Continental TSIO-520-C, still rated at 285 hp. Max cruising speed, weights and fuel are as above.

The U206B was approved Aug. 3, 1966. It has the Continental IO-520-F rated for five minutes of takeoff power at 300 hp. and 285 hp for other operations. Max cruising speed is still 170 mph., and weights and fuel are the same as above.

The P206C/TP206C was certificated July 20, 1967.The P206D/TP206D was certificated Sept. 18, 1968. The P206E/TP206E was approved July 28, 1969. All "P" models use the Continental IO-520-A while all "TP" models use the TSIO-520-C. Both engines are rated at 285 hp. Max cruising speed is still 170 mph. for all the models. Weights, fuel and oil remain the same.

The U206C/TU206C was approved July 20, 1967. The U206D/TU206D was approved Sept. 18, 1968. The U206E/TU206E was approved July 28, 1969. The U206F/TU206F was approved Oct. 26, 1971. Remember, the first model year is usually the year AFTER the certificate was approved.

The "U" models have the Continental IO-520-F rated at 300 hp. takeoff power, limited to five minutes, while the "TU" models have the Continental TSIO-520-C rated at 285 hp. Max structural cruising speed is still 170 mph, and weights, fuel and oil are about the same. Some serial numbers, however, (S/N U20602127 and on) had only 61 gallons, 59 usable.

The U206G/TU206G was certificated June 21, 1976. The "U" has the Continental IO-520-F engine with 300 hp. for five minutes at takeoff and 285 hp. thereafter. The "TU" has the Continental TSIO-520-M rated at **310 hp. for five minutes** and 285 hp. thereafter.

Fuel capacity is 61 gallons, 59 usable, with optional tanks offering either 80 gallons, 76 usable, or 92 gallons, 88 usable. Oil remains at 12 quarts. Weights for the U206 are 3,600 for landplane and 3,500 for floatplane. The TU206 has a max weight of 3,600, no matter if it is land, float or amphibian.

The 207 Type Certificate

The 207/T207 began life as the Skywagon/ Turbo Skywagon Dec. 31, 1968. It has the Continental IO-520-F engine rated at 300 hp. for five minutes and 285 hp. thereafter. It uses the McCauley D2A34C58/90AT-8 prop or the McCauley D3A32C90/82NC-2 prop. The T207 has the Continental TSIO-520-G with the same horsepower rating. Max structural cruising speed, like the 206, is 170 mph. Serial numbers 20700315 and up have a max structural cruising speed of 151 knots. It has seven seats, and carries either 65 gallons, 58 usable, or (from serial number 20700149 on up) it has 61 gallons, 54 usable. Max weight is up to 3,800 lbs. (It had better be, with seven people.)

The Eight-Seat Cessna

The 207A/T207A Skywagon/Turbo Skywagon; Stationair/Turbo Stationair was approved with seven seats July 12, 1976 for the 1977 model year and with EIGHT seats Sept. 11, 1979. The 207A has the Continental IO-520-F engine rated at 300 hp. for five minutes and 285 thereafter, while the T207A has the Continental TSIO-520-M rated at 310 hp. for five minutes, and 285 hp. thereafter.

Max structural cruising speed on the 207A is still 151 knots, while that of the T207A is 148 knots. Note: for serial numbers 2070483 and up, max structural cruising speed is 148 knots. Serial numbers 20700563 and up had eight seats.

Fuel capacity is 61 gallons, 54 usable, but optional tanks provide 80 gallons, 73 usable.

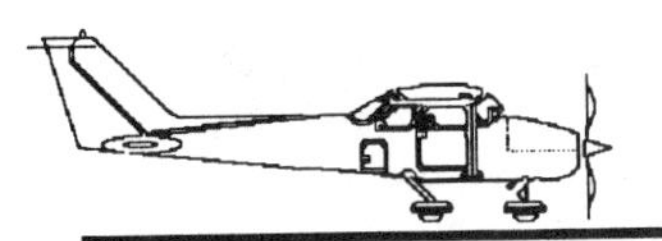
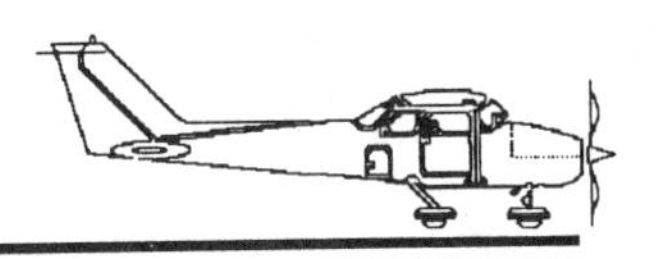

Notes at the end of the type certificates for the 207 are not unusual, and cover the normal weight and balance issues, placards and electrical systems.

A Look At Stationair Accidents

When the National Transportation Safety Board was asked for a computer printout of Cessna 206 accidents for 1986 and 1987, a rather thick pile of paper appeared in the mail. The aircraft has the Continental IO-520 engine. Let's take a look.

Feb. 7, 1986, Flagstaff, Az. Superstition Air Service, providing sightseeing for the Grand Canyon area in a TU-206G, landed off-airport when the exhaust system and stacks became loose. None of the eight people onboard was injured. The pilot noticed loss of power and the appearance of smoke under the instrument panel. The number one cylinder aft attach exhaust studs had sheared off, and the forward stud attach nuts were loose.

April 12, 1986, Wellsboro, Pa. The P206A lost power, landed in a field and hit a fence. The aircraft had heavy damage, but there were no injuries. The crankshaft had broken between the number two rod bearing journal and the number two main bearing journal. The engine had oil, but metal particles were present on both the oil screen and in the oil.

June 12, 1986, Tahoka, Texas. A 206 nosed over during a forced landing after fuel exhaustion. There was no fuel found anywhere in the fuel system after the nine-minute flight.

June 20, 1986, Ft. Myers, Fla. The nose gear wheel assembly separated from a U206F during landing. No annual inspection had been done for 18 months. During the one that WAS done, the wrong nuts had been installed on the wheel assembly.

July 13, 1986, Decker's Island, Texas. The pilot broke the main gear off on a fallen tree after failing to compensate for a crosswind.

Oct. 24, 1986, Greybull, Wy. Pilot left the fuel selector on the left tank, ran it dry, never switched to the full right tank, ran out of gas and stalled it into the ground, with no injuries.

Oct. 29, 1986, Marshfield, Mass. An alternator belt failed on a U206F, requiring the pilot to make a night, foggy, without-flaps landing. He came as close as humanly possible, landing in grass beside the runway. No injuries.

Dec. 6, 1986, Jean, Nev. A 29-year-old pilot with 2,930 hours ran both tanks dry while IFR with broken clouds at 400 feet, fog and light rain. He crashed into high terrain on the way to a highway to attempt a landing, killing a crew of two and four passengers. NTSB indicates there were serious injuries to one additional passenger. There were seven seats on the aircraft.

April 22, 1987, Temecula, Calif. The crankshaft failed from high cycle fatigue loading on a TU206B conducting an aerial survey flight. The fatigue loading started on the outside of one of the journals. Two people received minor injuries when the pilot was forced to switch his landing site from the runway to a nearby dirt road and nosed over.

June 1, 1987, Garrison, N.D. The number one engine cylinder of a TU206D had a catastrophic crack near the top, causing the engine to quit while photo-mapping at 3,000 ft. MSL. Two people received minor injuries when the nose-

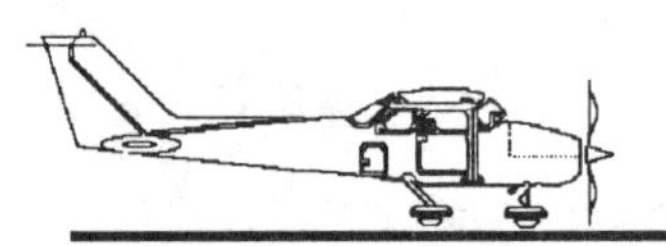
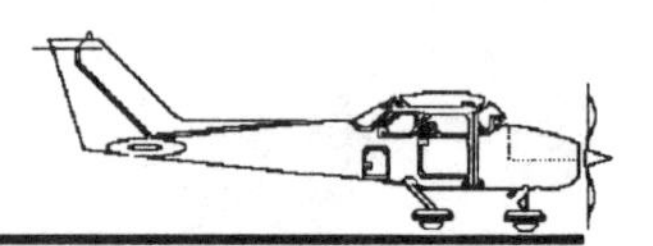

wheel dug in during a landing in a plowed field. The aircraft nosed over. The 27-year-old pilot had 2,290 hours.

July 4, 1987, Ft. Lauderdale, Fla. A 58-year-old pilot with 14,500 hours flew until the left tank went dry, then switched to the right but ran out of gas two miles from the destination. The right fuel cap was improperly seated, allowing fuel to siphon out. He landed in the clear zone of the Ft. Lauderdale Executive Airport and nosed over.

Sept. 16, 1987, Spanish Fork, Utah. A 24-year-old pilot with 1,800 hours was cruising at 10,500 ft. when the oil pressure dropped, the engine lost power and knocked, and a three-inch hole appeared in the top of the engine cowling. He attempted to land on a highway but decided to land beside it because there were too many cars on the road. He saw powerlines, pulled up and stalled it into the ground, with no injuries. The aircraft burned immediately. The engine teardown revealed a catastrophic failure of the number two piston and wrist pin. The aircraft had been overhauled June 5, 1984, and had flown 1,045 hours since.

Sept. 19, 1987, south of Kodiak, Alaska. A 206G amphibian suffered power loss and undershot a lake during the emergency landing. The number two bearing was found completely disintegrated and missing from its seat. There was evidence of high temperature and lack of oil lubrication.

Airworthiness Directives for the 206/207

Included here are airworthiness directives that are specific to the 206 or 207 and seem to be something you ought to know about. Many that apply to all Cessna models have been excluded, including the ones about poor seat locking and slippage of the shoulder harness.

July, 1986. To prevent engine power interruption due to loss of attachment of the engine controls. Applies to most Cessna models.

November, 1986. To prevent engine interruption due to contaminated fuel. Applies to many Cessna models.

January, 1984. To prevent power loss or engine stoppage due to water contamination of the fuel system. (Solution was to install quick drains. This AD applied to 182s, 206/207s and 210s.)

July, 1986. To preclude loss of the in-flight engine restart capability, and to correct a condition that could cause erroneous navigation displays to the pilot. (A voltage regulator was the problem.)

July, 1985. To prevent possible failure in the wing rear spar on airplanes that have had the full spar or inboard end of the spar replaced.

July, 1985. To eliminate the possibility of loss of the fuel selector roll pin installation.

January, 1985. To reduce the possibility of engine controls failure and loss of engine power control.

February, 1985. To eliminate the possibility of engine power reduction due to ingestion of pieces of a failed engine induction airbox outboard duct.

January, 1980. To preclude failure of the engine oil pressure and scavenge pump drive

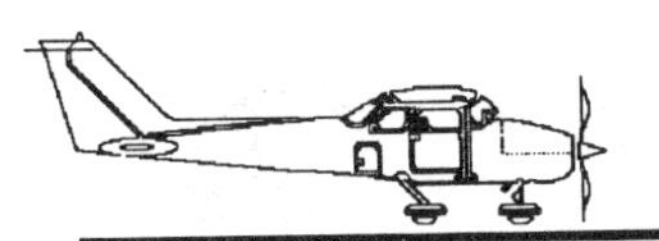

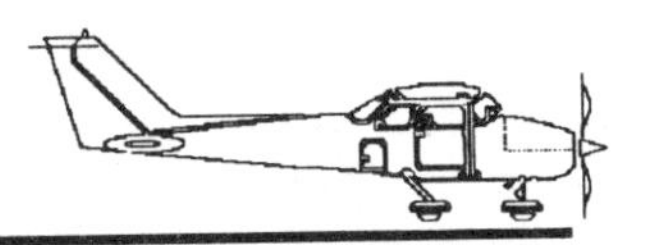

shaft and resulting oil pressure loss caused by turbocharger oil scavenge pump ingestion of failed turbocharger thrust bearing anti-rotation pins. Applies to TU206, TP206, T207, T210.

March, 1979. To preclude failure of the engine oil pressure and scavenge pump drive shaft and resulting oil pressure loss caused by turbocharger oil scavenge ingestion of failed turbocharger thrust bearing anti-rotation pins. Applies to TU206G, T210M and T207A. (See above.)

July, 1978 (78-07-01). To preclude engine oil pump failure due to contamination by the turbocharger thrust bearing anti-rotation pins, and failure of the turbocharger shaft.

January, 1979. To provide instructions for recognition of fuel system vapor blockage and operating procedures to restore normal fuel flow.

October, 1979. To provide an alternate source of fuel tank venting in case of fuel tank vent obstruction by foreign material and/or sticking of the fuel tank vent valve.

December, 1977. To prevent unwanted propeller rotation (when using external ground power receptacle). All this AD required was to apply power to the external ground power receptacle and see if the prop turns. If it doesn't, you have complied with the AD.

May, 1977. To prevent malfunction of the P/N C291503-0101 or P/N 1216100-1 fuel selector valve.

February, 1977. To replace a defective wing flap actuator ball nut assembly on the wing flap actuator P/N C301002-0101 (12-volt) or C301002-0102 (24-volt) date stamped OH, HH, WH or ZH.

March, 1976. To preclude restrictions of control movement due to jamming of the ARC PA-500A actuator gear train.

January, 1975. To preclude inadvertent fuel exhaustion due to incorrect fuel placarded capacities.

September, 1972. (75-07-09) To detect cracks and bolt looseness which could lead to in-flight separation of the fin and rudder. This applied to 205/206/210 airplanes with 1,000 hours of service, and required them to be rechecked at 1,000-hour intervals, except for floatplanes which are rechecked at 500-hour intervals.

July, 1971. To prevent exhaust gases from entering the cabin. (The exhaust manifold heat exchanger was blamed.)

April, 1971. To detect leakage of flammable fluids from flexible hose assemblies in the engine compartment.

March, 1972. To prevent inadvertent retraction of wing flaps and to insure positive operation of the electrical wing flap actuators.

Recurring airworthiness directives for the 200-series of Cessnas, which will add to your maintenance bills, include:

➤the 1987 seat locking AD,

➤the 1986 fuel contamination AD,

➤the 1985 engine power reduction due to ingestion of pieces of a failed engine AD,

➤the 1984 engine power loss caused by engine ingestion AD, [The engine that ate itself? Yes. See 206/207 accident listings.]

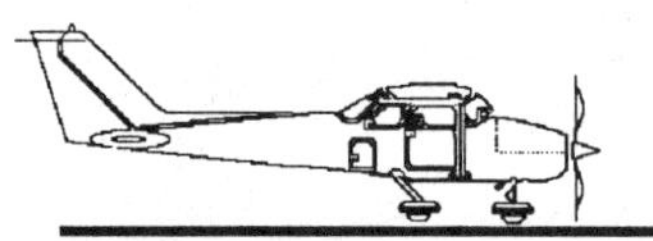
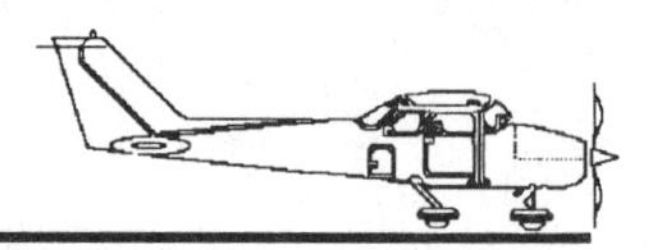

➤the 1982 Bendix loose distributor block bushing AD,

➤the 1981 failure of powerplant hoses carrying air, fuel and/or oil AD,

➤the 1978 McCauley propeller hub cracks AD,

➤the 1978 Bendix magneto impulse coupling failure AD,

➤the 1978 Goodyear fuel cells (bladders) AD,

➤the 1977 McCauley propeller hub cracks AD,

➤the 1976 main landing gear extension failure AD,

➤the 1976 Bendix inspection of ignition switches AD,

➤the 1974 Leigh Systems Sharc 7 emergency locator transmitter lithium battery AD,

➤the 1972 cracks and loose bolts in fin and rudder AD,

➤the 1972 inadvertent wing flap retraction AD,

➤the 1971 flexible hoses in engine compartment AD,

➤the 1968 Hartzell blade shank cracks AD, and

➤the 1967 Weston-Garwin Carruth loss of reliable altitude information AD.

It is a long, expensive list.

But then, this is meant to be a "nasty surprise" book, to better inform you of some of the bills ahead.

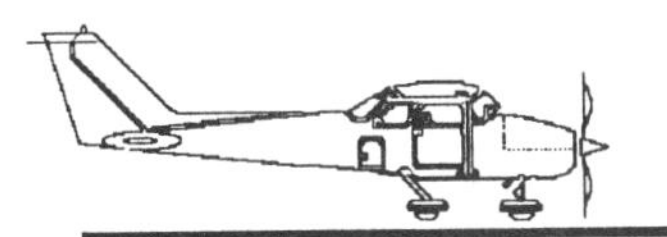
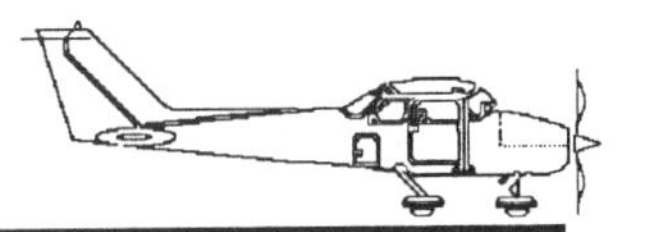

ENGINE OVERHAUL TERMS

SFNE Since factory new engine.

SFRM Since factory remanufactured engine was installed.

SMOH Since major overhaul.

STOH Since top overhaul, refers to cylinders, valves, pistons — everything from the base of the cylinder up. The bottom end refers to the crankcase, crankshaft, camshaft and bearings.

TBO Time between overhaul.

Borescope A device for seeing inside an engine to discover pitting, rust, etc.

Chromed cylinders Cylinder walls are ground oversize, and then treated with chrome and honed back to standard size. Chrome offers corrosion and wear protection.

Compression The pressure in pounds per square inch inside the cylinder when the piston is at the very top of its travel on the compression stroke. Eighty psi of air is injected into the cylinder, while a gauge measures the pressure that is retained.

Galling Wear between moving parts.

New limits Parts are the same dimensions as new parts. Parts that meet service limits show wear but are sufficient for service.

Nitride A process that hardens the surface of steel, typically applied to crankshaft bearing surfaces and cylinder walls in reciprocating engines.

APPENDICES

U.S. Department
of Transportation

Federal Aviation Administration

AC 20-5F
Revised
1986

Plane Sense

GENERAL AVIATION INFORMATION

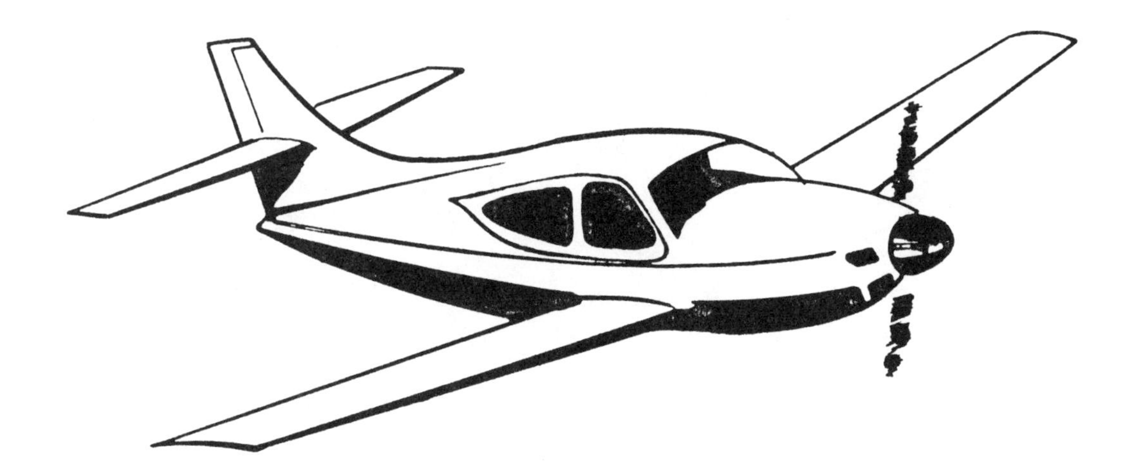

PREFACE

Plane Sense was prepared by the U.S. Department of Transportation, Federal Aviation Administration, Office of Flight Standards, to acquaint the prospective owner with some fundamental information on the requirements of owning and operating a private airplane.

Anyone who is seriously thinking of becoming an aircraft owner should become familiar with the Federal Aviation Regulations. Since the aviation picture is constantly changing, it is suggested that you contact your nearest FAA General Aviation or Flight Standards District Office, where the personnel will be glad to acquaint you with the latest requirements of private ownership.

Comments regarding this publication should be directed to:

U.S. Department of Transportation
Federal Aviation Administration
Aviation Standards National Field Office
Examinations Standards Branch, AVN–130
P.O. Box 25082
Oklahoma City, OK 73125

This advisory circular supersedes AC 20–5E, dated 1981.

CONTENTS

PLANE SENSE

General Aviation Information

Revised

1986

U.S. DEPARTMENT OF TRANSPORTATION
FEDERAL AVIATION ADMINISTRATION
Office of Flight Standards

BUYING AN AIRCRAFT

When buying a used aircraft, it is wise to have the selected aircraft inspected by a qualified person or facility before you buy. The condition of the aircraft and the state of its maintenance records can be determined by persons familiar with the particular make and model. These include an FAA certificated A & P (airframe and powerplant) mechanic or an approved repair station.

Q. What is meant by a *clear title*?

A. A *clear title* is a term commonly used by aircraft title search companies to indicate there are no liens (chattel mortgage, security agreement, tax lien, artisan lien, etc.) in the FAA (Federal Aviation Administration) aircraft records. The FAA Aircraft Registry does not perform title searches for the aviation public; however, the aircraft record contains all of the ownership and security documents which have been filed with the FAA.

Q. How can I be sure that the aircraft has a *clear title*?

A. Either search the aircraft records yourself, or have it done by an attorney or qualified aircraft title search company. A list of title search companies qualified in aircraft title and records search can be found on AC Form 8050–55, Title Search Companies.

You wouldn't think of purchasing a house until you had the records examined. You should do no less when purchasing an aircraft, which also represents a substantial investment. Even though you are planning to purchase the aircraft from an established dealer, it makes good sense to determine the true status of the aircraft records before you buy. CAUTION: FAA registration cannot be used in any civil proceedings to establish proof of ownership!

There is no substitute for examining the aircraft records to secure a history of the ownership of the aircraft and to determine if there are any outstanding liens or mortgages. This procedure will help avoid a delay in registering an aircraft and the headaches many have suffered because they failed to take this one important step before purchasing their aircraft.

Q. Where do I go to search the records?

A. Aircraft records maintained by the FAA are on file at the Mike Monroney Aeronautical Center, Aviation Records Building, Aircraft Registry, AAC–250, 6500 South MacArthur Boulevard, P.O. Box 25504, Oklahoma City, OK 73125 (telephone (405) 686–2116). Records may be requested and reviewed at this address. There may be other records filed at state/local level which are not recorded with the FAA.

Q. What documents may I expect to receive with my new or used aircraft?

A. 1—Bill of sale or conditional sales contract.
2—Either FAA Form 8100-2, Standard Airworthiness Certificate, or FAA Form 8130-7, Special Airworthiness Certificate.
3—Maintenance records containing the following information:
(a) The total time in service of the airframe, each engine, and each propeller;
(b) The current status of life-limited parts of each airframe, engine, propeller, rotor, and appliance;
(c) The time since last overhaul of all items installed on the aircraft that are required to be overhauled on a specified time basis;
(d) The identification of the current inspection status of the aircraft, including the times since the last inspections required by the inspection program under which the aircraft and its appliances are maintained;
(e) The current status of applicable AD's (Airworthiness Directives) including, for each, the method of compliance, the AD number, and revision date. If the AD involves recurring action, the time and date when the next action is required; and
(f) A copy of current major alterations to each airframe, engine, propeller, rotor, and appliance.
4—Equipment list and weight and balance data.
5—Airplane Flight Manual or operating limitations.

Q. What manuals should I receive with the aircraft?

A. The manufacturers produce owner's manuals, maintenance manuals, service letters and bulletins, and other technical data pertaining to their aircraft. These may be available from the previous owner, but are not part of the aircraft and are not required to be transferred to a new owner as are the aforementioned five items. If the service manuals are not available from the previous owner, they usually may be obtained from the aircraft manufacturer.

Q. What is the meaning of *airworthy*?

A. Two conditions must be met for an aircraft to be considered *airworthy*. These conditions are:

1—The aircraft conforms to its type design (type certificate). Conformity to type design is considered attained when the required and proper components are installed and they are consistent with the drawings, specifications, and other data that are a part of the type certificate. Conformity would include applicable supplemental type certificates and field–approved alterations.

2—The aircraft is in condition for safe operation. This refers to the condition of the aircraft with relation to wear and deterioration.

Q. Does a current 100–hour or annual inspection mean that the aircraft is in *first class* condition?

A. No. It indicates only that the aircraft was found to be in airworthy condition at the time of inspection.

Q. What should I do before buying an amateur–built or experimental aircraft?

A. 1—Examine the Airworthiness Certificate and its operating limitations.
   2—Contact the General Aviation or Flight Standards District Office serving your locale and ask to speak to an airworthiness inspector who will explain the requirements for experimental certification.

Q. What should I consider when buying a surplus military aircraft?

A. Certain surplus military aircraft are not eligible for FAA certification in the STANDARD, RESTRICTED, or LIMITED classifications. Since no civil aircraft may be flown unless certificated, you should discuss this with the local FAA inspector, who will advise you of eligible aircraft and certification procedures.

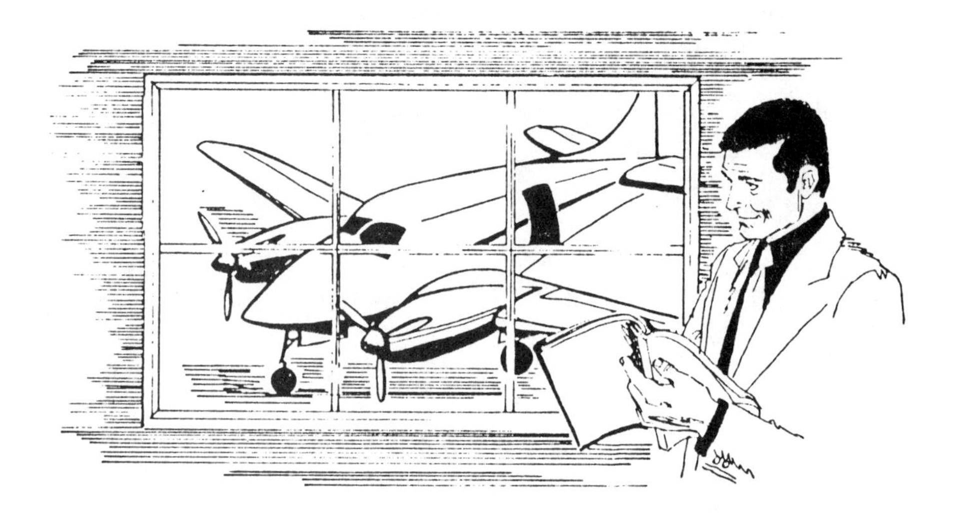

AIRCRAFT OWNER RESPONSIBILITIES

You, as an aircraft owner, will be assuming responsibilities similar to those you have if you own an automobile. Owning an automobile usually means that you must register it in your state of residence and obtain license plates. As the registered owner of an aircraft, you will be responsible for:

1—Having a current Airworthiness Certificate and Certificate of Aircraft Registration in your aircraft.
2—Maintaining your aircraft in an airworthy condition including compliance with all applicable AD's.
3—Assuring that maintenance is properly recorded.
4—Keeping abreast of current regulations concerning the operation and maintenance of your aircraft.
5—Notifying the FAA Aircraft Registry immediately of any change of permanent mailing address, of the sale or export of your aircraft, or of the loss of your eligibility to register an aircraft. (See FAR (Federal Aviation Regulation) Section 47.41.)

Some states require that your automobile be inspected periodically to assure that it is in safe operating condition. Your aircraft will have to be inspected in accordance with an annual inspection or with one of the inspection programs outlined in FAR Section 91.169, in order to maintain a current Airworthiness Certificate. As with your automobile, accidents involving your aircraft must also be reported.

Some similarities between automobile and aircraft responsibilities are shown in the following chart:

Automobile/Airplane Comparison Chart

Responsibility	*Automobile*	*Aircraft*
Registration	Yes	Yes
Inspection	Yes	Yes
Compulsory insurance (most states)	Yes	No
Reporting of accidents	Yes	Yes
Required maintenance records	No	Yes
Maximum speed restrictions	Yes	Yes
Controlled maintenance	No	Yes

How to Report A Stolen Aircraft

1—Immediately notify the law enforcement agency having jurisdiction at the site of the theft, giving all available information. Request that such information be entered into the computer system of the National Crime Information Center of the FBI, and have the law officer taking the report notify the nearest FAA Flight Service Station. The Flight Service Station, in turn, will issue a nationwide stolen aircraft alert. NOTE: Flight Service Stations are prohibited from issuing stolen aircraft alerts based solely on notification of theft by the owner—the report must be made by the law enforcement officer handling the case.

2—Notify the International Aviation Theft Bureau—telephone (301) 654-0500; TELEX 89-8468; TWX 710-824-0095—giving all available information to activate the United States Customs Service and law enforcement alerting network of Mexico.

3—Notify your insurance company or agent, as appropriate.

Additionally, owners/operators are encouraged to keep separate records of engine and equipment serial numbers, and report these serial numbers at the same time the stolen aircraft is reported.

AIRCRAFT REGISTRATION

Eligible Registrants

An aircraft is eligible for registration in the United States only if it is owned by:

1—A U.S. citizen (individual, partnership, or corporation);
2—A resident alien;
3—A corporation (other than one which is a U.S. citizen), lawfully organized and doing business under the laws of the United States or of any state thereof, if the aircraft is based and used primarily in the United States; or
4—A government entity (federal, state, or local).

The aircraft may not be registered in a foreign country during the period it is registered in the United States.

If you purchase an aircraft, you must apply for a Certificate of Aircraft Registration from the FAA Aircraft Registry before it may be operated. Do not depend on a bank, loan company, aircraft dealer, or anyone else to submit the application for registration. Do it yourself (in the name of the owner, not in the name of the bank or other mortgage holder).

Aircraft Previously Registered in the United States

You should immediately submit evidence of ownership, an Aircraft Registration Application, and a $5 registration fee to the Federal Aviation Administration, Aircraft Registry, AAC–250, P.O. Box 25504, Oklahoma City, OK 73125. Fees required for aircraft registration may be paid by check or money order made payable to the Treasurer of the United States.

A bill of sale form that meets the FAA's requirements for evidence of ownership is AC Form 8050–2, Aircraft Bill of Sale, which may be obtained from the nearest FAA district office. The form includes an information and instruction sheet. If a conditional sales contract is the evidence of ownership, an additional $5 fee is required for recording. For FAA registration, the bill of sale need not be notarized. (See Figure 1.)

The Aircraft Registration Application includes an information and instruction sheet. Submit the white and green copies to the FAA Aircraft Registry; keep the pink copy in your aircraft as evidence of application for registration until you receive your Certificate of Aircraft Registration. The pink copy is good for 90 days. Registration certificates are issued to the person whose name is on the application. (See Figure 2, page 12.)

If there is a break in the chain of ownership of the aircraft, i.e., if it is not being purchased from the last registered owner, you are required to submit conveyances to complete the chain of ownership, through all intervening owners, including yourself, to the FAA Aircraft Registry.

The Aircraft Registration Application may also be used to report a change of address by the aircraft owner. The FAA will issue a revised certificate at no charge.

FORM APPROVED
OMB No 2120-0029
EXP DATE 10/31/84

UNITED STATES OF AMERICA
DEPARTMENT OF TRANSPORTATION

AIRCRAFT BILL OF SALE

FOR AND IN CONSIDERATION OF $ THE UNDERSIGNED OWNER(S) OF THE FULL LEGAL AND BENEFICIAL TITLE OF THE AIRCRAFT DESCRIBED AS FOLLOWS:

UNITED STATES REGISTRATION NUMBER N 123BJ

AIRCRAFT MANUFACTURER & MODEL
BIG DEAL BA-OH

AIRCRAFT SERIAL No.
00021

DOES THIS 25th DAY OF July 19 85
HEREBY SELL, GRANT, TRANSFER AND DELIVER ALL RIGHTS, TITLE, AND INTERESTS IN AND TO SUCH AIRCRAFT UNTO:

Do Not Write In This Block
FOR FAA USE ONLY

PURCHASER

NAME AND ADDRESS
(IF INDIVIDUAL(S), GIVE LAST NAME, FIRST NAME, AND MIDDLE INITIAL.)

WILLIAMS, MARION W.
1000 Whitehouse Road
Oklahoma City, OK 73100

DEALER CERTIFICATE NUMBER D112762

AND TO HIS EXECUTORS, ADMINISTRATORS, AND ASSIGNS TO HAVE AND TO HOLD SINGULARLY THE SAID AIRCRAFT FOREVER, AND WARRANTS THE TITLE THEREOF.

IN TESTIMONY WHEREOF I HAVE SET MY HAND AND SEAL THIS 25 DAY OF Jul 19 85

SELLER

NAME (S) OF SELLER (TYPED OR PRINTED)	SIGNATURE (S) (IN INK) (IF EXECUTED FOR CO-OWNERSHIP, ALL MUST SIGN.)	TITLE (TYPED OR PRINTED)
JONES SAFETY CORPORATION	John B. Jones	JOHN B. JONES President

ACKNOWLEDGMENT (NOT REQUIRED FOR PURPOSES OF FAA RECORDING: HOWEVER, MAY BE REQUIRED BY LOCAL LAW FOR VALIDITY OF THE INSTRUMENT.)

ORIGINAL: TO FAA

AC FORM 8050-2 (9-82) (0052-00-629-0002)

Figure 1. AC Form 8050–2, Aircraft Bill of Sale

FORM APPROVED
OMB NO 2120-0029
EXP DATE 10/31/84

UNITED STATES OF AMERICA DEPARTMENT OF TRANSPORTATION
FEDERAL AVIATION ADMINISTRATION-MIKE MONRONEY AERONAUTICAL CENTER
AIRCRAFT REGISTRATION APPLICATION

CERT. ISSUE DATE

UNITED STATES REGISTRATION NUMBER **N** 123BJ

AIRCRAFT MANUFACTURER & MODEL
BIG DEAL BA-OH

AIRCRAFT SERIAL No
00021

FOR FAA USE ONLY

TYPE OF REGISTRATION (Check one box)

[X] 1 Individual [ ] 2 Partnership [ ] 3 Corporation [ ] 4 Co-owner [ ] 5 Gov't [ ] 8 Foreign-owned Corporation

NAME OF APPLICANT (Person(s) shown on evidence of ownership If individual, give last name, first name, and middle initial)

WILLIAMS, MARION W.

TELEPHONE NUMBER ( )

ADDRESS (Permanent mailing address for first applicant listed)

Number and street 1000 Whitehouse Road

Rural Route P O Box

CITY	STATE	ZIP CODE
Oklahóma City	OK	73100

[ ] **CHECK HERE IF YOU ARE ONLY REPORTING A CHANGE OF ADDRESS**

ATTENTION! Read the following statement before signing this application.

A false or dishonest answer to any question in this application may be grounds for punishment by fine and / or imprisonment (U S Code, Title 18, Sec 1001)

CERTIFICATION

I/WE CERTIFY

(1) That the above aircraft is owned by the undersigned applicant, who is a citizen (including corporations) of the United States

(For voting trust, give name of trustee ______________________) or

CHECK ONE AS APPROPRIATE

a [ ] A resident alien, with alien registration (Form 1-151 or Form 1-551) No ______________

b [ ] A foreign-owned corporation organized and doing business under the laws of (state or possession) ______________, and said aircraft is based and primarily used in the United States. Records of flight hours are available for inspection at ______________

(2) That the aircraft is not registered under the laws of any foreign country, and
(3) That legal evidence of ownership is attached or has been filed with the Federal Aviation Administration

NOTE If executed for co-ownership all applicants must sign Use reverse side if necessary

TYPE OR PRINT NAME BELOW SIGNATURE

EACH PART OF THIS APPLICATION MUST BE SIGNED IN INK	SIGNATURE	TITLE	DATE
	Marion W. Williams	Owner	7/25/85
	SIGNATURE	TITLE	DATE
	SIGNATURE	TITLE	DATE

NOTE Pending receipt of the Certificate of Aircraft Registration, the aircraft may be operated for a period not in excess of 90 days, during which time the PINK copy of this application must be carried in the aircraft

AC FORM 8050-1 (1-83) (0052-00-628-9005)

Figure 2. AC Form 8050–1, Aircraft Registration Application

If the certificate is lost, destroyed, or mutilated, a replacement may be obtained at the written or telegraphic request of the holder. Send the request and $2 (check or money order payable to the Treasurer of the United States) to:

Federal Aviation Administration
Aircraft Registry, AAC–250
P.O. Box 25504
Oklahoma City, OK 73125

The request should describe the aircraft by make, model, serial number, and registration number. If operation of the aircraft is necessary before receipt of the duplicate certificate, the FAA Aircraft Registry will, if requested, send telegraphic authority (collect) upon receipt of the $2 duplicate certificate fee. Include in your request your full address and a telex number, if available, to which the telegram may be charged. A telephone number where you can be reached should be included.

Aircraft Previously Registered in a Foreign Country

If you are contemplating purchase of an aircraft registered in a foreign country, contact the local FAA district office for certification assistance and the FAA Aircraft Registry at (405) 686–2116 for registration assistance.

Certificate of Aircraft Registration

A Certificate of Aircraft Registration should be in the aircraft before an Airworthiness Certificate can be issued. The Certificate of Aircraft Registration will expire as described in FAR Section 47.41 when: (See Figure 3, page 14.)

1—The aircraft becomes registered under laws of a foreign country;
2—The registration of the aircraft is canceled at the written request of the holder of the certificate;
3—The aircraft is totally destroyed or scrapped;
4—The holder of the certificate loses his or her U.S. citizenship or status as an alien (without becoming a U.S. citizen);
5—The ownership of the aircraft is transferred; or
6—Thirty days have elapsed since the death of the holder of the certificate.

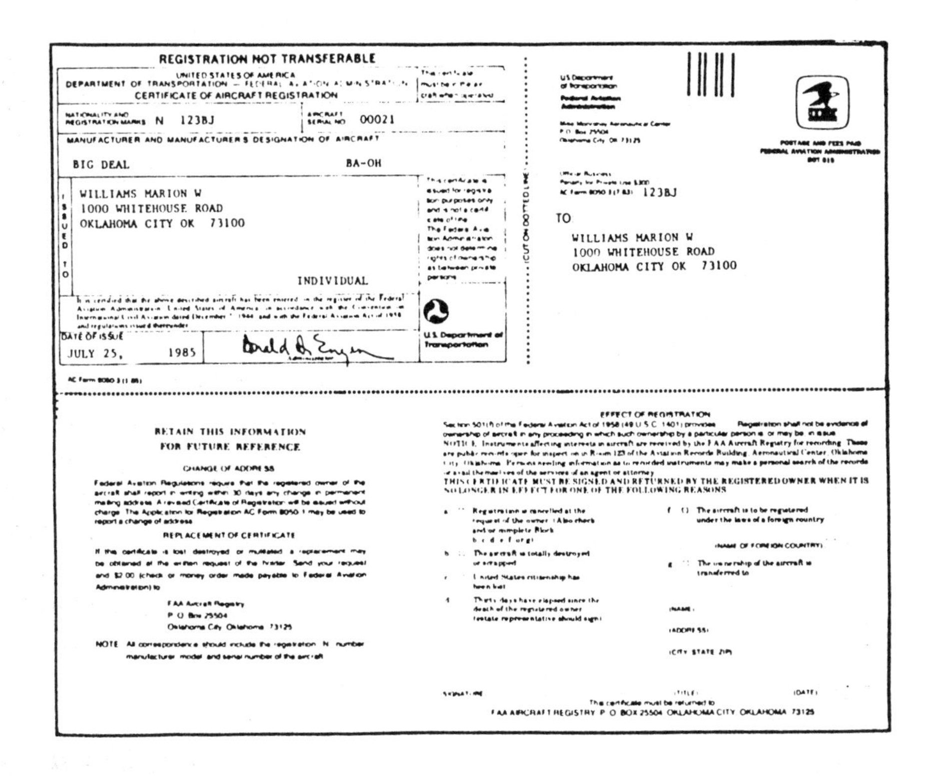

REGISTRATION NOT TRANSFERABLE

UNITED STATES OF AMERICA
DEPARTMENT OF TRANSPORTATION — FEDERAL AVIATION ADMINISTRATION
CERTIFICATE OF AIRCRAFT REGISTRATION

NATIONALITY AND REGISTRATION MARKS N 123BJ

AIRCRAFT SERIAL NO. 00021

MANUFACTURER AND MANUFACTURER'S DESIGNATION OF AIRCRAFT

BIG DEAL BA-OH

ISSUED TO

WILLIAMS MARION W
1000 WHITEHOUSE ROAD
OKLAHOMA CITY OK 73100

INDIVIDUAL

This certificate is issued for registration purposes only and is not a certificate of title. The Federal Aviation Administration does not determine rights of ownership as between private persons.

DATE OF ISSUE

JULY 25, 1985

U.S. Department of Transportation

AC Form 8050-3 (1-85)

(CUT ON DOTTED LINE)

US Department of Transportation
Federal Aviation Administration

P.O. Box 25504
Oklahoma City OK 73125

POSTAGE AND FEES PAID
FEDERAL AVIATION ADMINISTRATION
DOT 515

Official Business
Penalty for Private Use $300
123BJ

TO

WILLIAMS MARION W
1000 WHITEHOUSE ROAD
OKLAHOMA CITY OK 73100

RETAIN THIS INFORMATION
FOR FUTURE REFERENCE

CHANGE OF ADDRESS

Federal Aviation Regulations require that the registered owner of the aircraft shall report in writing within 30 days any change in permanent mailing address. A revised Certificate of Registration will be issued without charge. The Application for Registration AC Form 8050-1 may be used to report a change of address.

REPLACEMENT OF CERTIFICATE

If the certificate is lost, destroyed or mutilated, a replacement may be obtained at the written request of the holder. Send your request and $2.00 (check or money order made payable to Federal Aviation Administration) to:

FAA Aircraft Registry
P.O. Box 25504
Oklahoma City, Oklahoma 73125

NOTE: All correspondence should include the registration N-number, manufacturer, model and serial number of the aircraft.

EFFECT OF REGISTRATION

Section 501(f) of the Federal Aviation Act of 1958 (49 U.S.C. 1401) provides: "Registration shall not be evidence of ownership of aircraft in any proceeding in which such ownership by a particular person is, or may be, in issue."
NOTICE: Instruments affecting interests in aircraft are received by the FAA Aircraft Registry for recording. These are public records open for inspection in Room 123 of the Aviation Records Building, Aeronautical Center, Oklahoma City, Oklahoma. Persons needing information as to recorded instruments may make a personal search of the records or avail themselves of the services of an agent or attorney.
THIS CERTIFICATE MUST BE SIGNED AND RETURNED BY THE REGISTERED OWNER WHEN IT IS NO LONGER IN EFFECT FOR ONE OF THE FOLLOWING REASONS:

a. Registration is cancelled at the request of the owner. (Also check and/or complete Block b, c, d, e, f or g.)
b. The aircraft is totally destroyed or scrapped.
c. United States citizenship has been lost.
d. Thirty days have elapsed since the death of the registered owner. (estate representative should sign)
f. ( ) The aircraft is to be registered under the laws of a foreign country.

(NAME OF FOREIGN COUNTRY)

g. The ownership of the aircraft is transferred to:

(NAME)

(ADDRESS)

(CITY, STATE, ZIP)

SIGNATURE (TITLE) (DATE)

This certificate must be returned to:
FAA AIRCRAFT REGISTRY, P.O. BOX 25504, OKLAHOMA CITY, OKLAHOMA 73125

Figure 3. AC Form 8050–3, Certificate of Aircraft Registration

When an aircraft is destroyed, scrapped, or sold, the owner shall notify the FAA by filling in the back of the Certificate of Aircraft Registration and mailing it to:

Federal Aviation Administration
Aircraft Registry, AAC–250
P.O. Box 25504
Oklahoma City, OK 73125

The U.S. registration and nationality marking should be removed from an aircraft before it is delivered to a purchaser who is not eligible to register it in the United States. The endorsed Certificate of Aircraft Registration should be forwarded to the FAA Aircraft Registry.

A dealer's aircraft registration certificate is another form of registration. It is valid only for flights within the United States by the manufacturer or a dealer for flight testing or demonstration for sale. It should be removed by the dealer when the aircraft is sold.

The certificate of registration serves as conclusive evidence of nationality but is not a title and is not evidence of ownership in any proceeding in which ownership is in issue.

Special Registration Number (N – Number)

A U.S. identification number of your choice may be reserved, if available. This number may not exceed five characters in addition to the prefix letter "N," and may be one to five numbers (N11111), one to four numbers and one suffix letter (N1000A), or one to three numbers and two suffix letters (N100AA).

In your written request, list up to five numbers in order of preference in the event the first choice is not available; also include a $10 fee. If your request is approved, you will be notified that the number has been reserved for 1 year. You will also be informed that this reservation may be extended on a yearly basis for a $10 renewal fee.

When you are ready to place the number on your aircraft, you should request permission by forwarding a complete description of the aircraft to the FAA Aircraft Registry. Permission to place the special number on your aircraft will be given on AC Form 8050–64, Assignment of Special Registration Numbers. When the number is placed on your aircraft, sign and return the original to the FAA Aircraft Registry within 5 days. (See Figure 4.)

The duplicate of AC Form 8050–64, together with your Airworthiness Certificate, should then be presented to an FAA inspector, who will issue a revised Airworthiness Certificate showing the new N – Number. The old registration certificate and the duplicate AC Form 8050–64 should be carried in the aircraft until the new registration certificate is received.

Registration of Amateur–Built Aircraft

AC Form 8050–88, Identification Number Assignment and Registration of Amateur–Built Aircraft, is used by the FAA Aircraft Registry to notify you of action taken on your application for registration of amateur–built aircraft. The reverse side of AC Form 8050–88 is an Affidavit of Ownership for an amateur–built aircraft. It is completed when applying for registration of an amateur–built aircraft. (See Figures 5 and 6, pages 18 and 19.)

Additional Information

FAR Part 47 specifies the requirement for registering aircraft. For information concerning FAR Part 47 or any circumstances not discussed herein, contact the Federal Aviation Administration, Aircraft Registry, AAC–250, P.O. Box 25504, Oklahoma City, OK 73125. Telephone (405) 686–2116 for registration information and (405) 686–4206 for N – Number information.

State registration of aircraft is required in approximately 60 percent of the states. Check for your state's requirement.

US Department of Transportation
Federal Aviation Administration

ASSIGNMENT OF SPECIAL REGISTRATION NUMBERS

Special Registration Number: N 7316

Aircraft Make and Model: BIG DEAL, BA-OH

Present Registration Number: N 123BJ

Serial Number: 00021

Issue Date: July 31, 1985

MARION W. WILLIAMS
1000 Whitehouse Road
Oklahoma City, OK 73100

This is your authority to change the United States registration number on the above described aircraft to the special registration number shown

Carry duplicate of this form in the aircraft together with the old registration certificate as interim authority to operate the aircraft pending receipt of revised certificate of registration. Obtain a revised certificate of airworthiness from your nearest Flight Standards field office

The latest FAA Form 8130 6 on file is dated

The airworthiness classification and category

SIGN AND RETURN THE ORIGINAL of this form to the FAA Aircraft Registry within 5 days after placing the special registration number on the aircraft. A revised certificate will then be issued. Unless this authority is used and this office so notified, the authority for use of the special number will expire on

CERTIFICATION I certify that the special registration number was placed on the aircraft described above

Sign of Owner: Marion W. Williams

Title of Owner:

Date Placed on Aircraft: August 1, 1985

RETURN FORM TO

FAA Aircraft Registry
P.O. Box 25504
Oklahoma City, Oklahoma 73125

BELOW THIS POINT FOR FAA USE ONLY

FP NF NAME

ADDRESS

FC ZIP EMP CODE DATE

AC Form 8050-64 (11-82)

Figure 4. AC Form 8050–64, Assignment of Special Registration Numbers

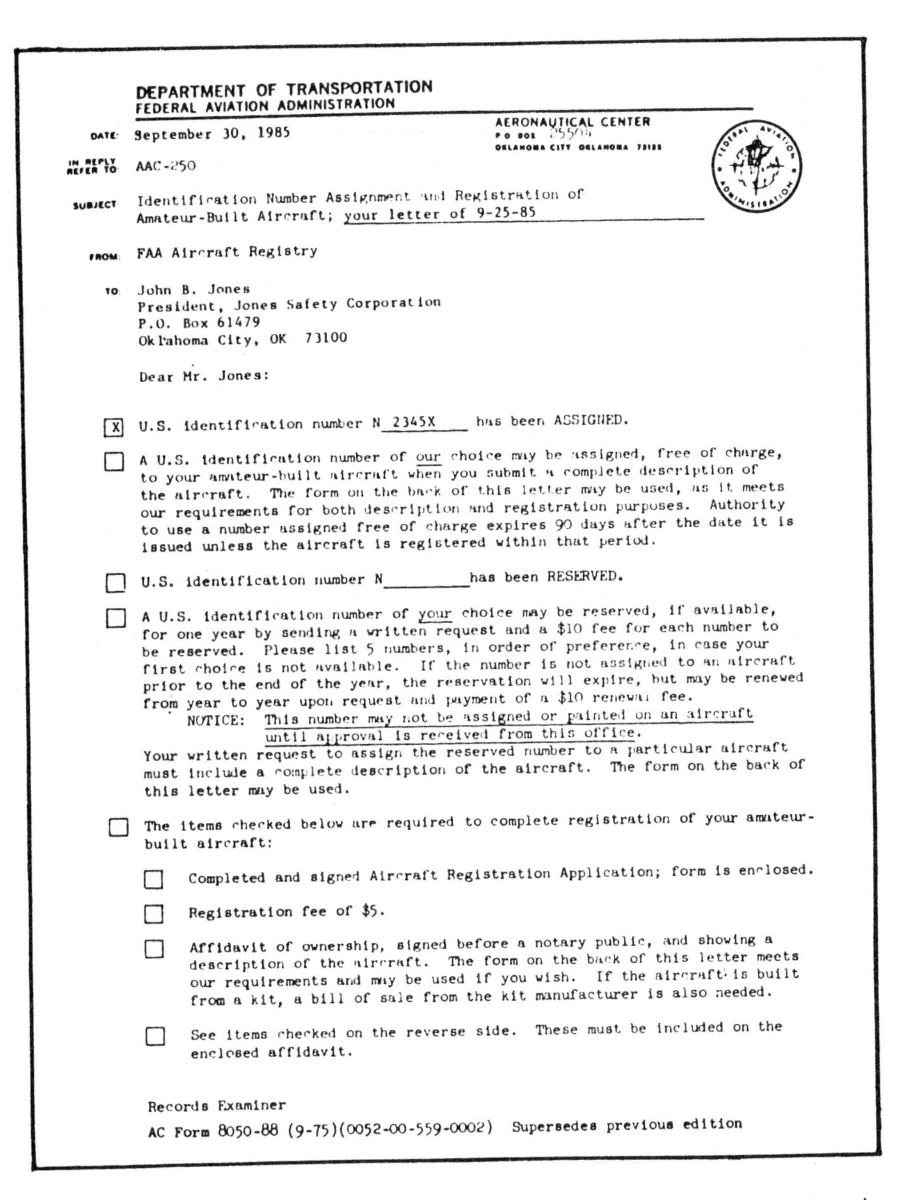

DEPARTMENT OF TRANSPORTATION
FEDERAL AVIATION ADMINISTRATION

AERONAUTICAL CENTER
P.O. BOX 25504
OKLAHOMA CITY, OKLAHOMA 73125

DATE: September 30, 1985

IN REPLY REFER TO: AAC-250

SUBJECT: Identification Number Assignment and Registration of Amateur-Built Aircraft; your letter of 9-25-85

FROM: FAA Aircraft Registry

TO: John B. Jones
President, Jones Safety Corporation
P.O. Box 61479
Oklahoma City, OK 73100

Dear Mr. Jones:

[X] U.S. identification number N 2345X has been ASSIGNED.

[ ] A U.S. identification number of our choice may be assigned, free of charge, to your amateur-built aircraft when you submit a complete description of the aircraft. The form on the back of this letter may be used, as it meets our requirements for both description and registration purposes. Authority to use a number assigned free of charge expires 90 days after the date it is issued unless the aircraft is registered within that period.

[ ] U.S. identification number N__________ has been RESERVED.

[ ] A U.S. identification number of your choice may be reserved, if available, for one year by sending a written request and a $10 fee for each number to be reserved. Please list 5 numbers, in order of preference, in case your first choice is not available. If the number is not assigned to an aircraft prior to the end of the year, the reservation will expire, but may be renewed from year to year upon request and payment of a $10 renewal fee.
NOTICE: This number may not be assigned or painted on an aircraft until approval is received from this office.
Your written request to assign the reserved number to a particular aircraft must include a complete description of the aircraft. The form on the back of this letter may be used.

[ ] The items checked below are required to complete registration of your amateur-built aircraft:

- [ ] Completed and signed Aircraft Registration Application; form is enclosed.
- [ ] Registration fee of $5.
- [ ] Affidavit of ownership, signed before a notary public, and showing a description of the aircraft. The form on the back of this letter meets our requirements and may be used if you wish. If the aircraft is built from a kit, a bill of sale from the kit manufacturer is also needed.
- [ ] See items checked on the reverse side. These must be included on the enclosed affidavit.

Records Examiner

AC Form 8050-88 (9-75)(0052-00-559-0002) Supersedes previous edition

Figure 5. AC Form 8050–88, Identification Number Assignment and Registration of Amateur–Built Aircraft

AFFIDAVIT OF OWNERSHIP FOR AMATEUR-BUILT AIRCRAFT

U.S. Identification Number N2345X

Builder's Name Charles E. Griffin

Model CFG-1 Serial Number 00001

Class (airplane, rotorcraft, glider, etc.) airplane

Type of Engine Installed (reciprocating, turbopropeller, etc.)

turbopropeller

Number of Engines Installed 1

Manufacturer, Model, and Serial Number of each Engine Installed

Twister, PHP, 5064

Built for Land or Water Operation land

Number of Seats 1

The above-described aircraft was built from parts by the undersigned and I am the owner.

Charles E. Griffin
(Signature of Owner-Builder)

State of Oklahoma

County of Oklahoma

Subscribed and sworn to before me this 23 day of September, 19 85.

My commission expires 12/16/87.

A. B. Jackson
(Signature of Notary Public)

AC Form 8050-88 (9-75) (0052-00-559-0002) Supersedes previous edition

Figure 6. AC Form 8050–88, Identification Number Assignment and Registration of Amateur–Built Aircraft, (reverse side) Affidavit of Ownership for Amateur–Built Aircraft

AIRWORTHINESS CERTIFICATE

An Airworthiness Certificate is issued by a representative of the FAA after the aircraft has been inspected, is found to meet the requirements of the FAR's, and is in condition for safe operation. The certificate must be displayed in the aircraft so that it is legible to passengers or crew whenever the aircraft is operated. The Airworthiness Certificate is transferred with the aircraft, except when it is sold to a foreign purchaser.

The FAA Form 8100–2, Standard Airworthiness Certificate, is issued for aircraft type certificated in the normal, utility, acrobatic, and transport categories, or for manned free balloons. An explanation of each term in the certificate follows: (See Figure 7, page 23.)

Item 1. Nationality—The "N" indicates the aircraft is of U.S. registry. Registration Marks—the number, in this case 12345, is the registration number assigned to the aircraft.

Item 2. Indicates the make and model of the aircraft.

Item 3. Is the manufacturer's serial number assigned to the aircraft, as noted on the aircraft data plate.

Item 4. Indicates that the aircraft, in this case, must be operated in accordance with the limitations specified for the NORMAL category.

Item 5. Indicates the aircraft has been found to conform to its type certificate and is considered in condition for safe operation at the time of inspection and issuance of the certificate. Any exemptions from the applicable airworthiness standards are briefly noted here and the exemption number given. The word NONE will be entered if no exemption exists.

Item 6. Indicates the Airworthiness Certificate is in effect indefinitely, if the aircraft is maintained in accordance with FAR Parts 21, 43, and 91, and the aircraft is registered in the United States. Also included are the date the certificate was issued, the signature of the FAA representative, and his or her office identification.

A Standard Airworthiness Certificate remains in effect as long as the aircraft receives the required maintenance and is properly registered in the United States. Flight safety relies in part on the condition of the aircraft, which may be determined on inspection by mechanics, approved repair stations, or manufacturers who meet specific requirements of FAR Part 43.

The FAA Form 8130–7, Special Airworthiness Certificate, is issued for all aircraft certificated in other than the Standard classifications (Experimental, Restricted, Limited, and Provisional). (See Figure 8.)

If you are interested in purchasing an aircraft classed as other than Standard, it is suggested that you contact the local FAA General Aviation or Flight Standards District Office for an explanation of the pertinent airworthiness requirements and the limitations of such a certificate.

In summary, the FAA initially determines that your aircraft is in condition for safe operation and conforms to type design, then issues an Airworthiness Certificate.

Advisory Circular 21–12, Application for U.S. Airworthiness Certificate, (see Figure 9, page 24, FAA Form 8130–6) provides additional information if needed. (See page 41 for ordering instructions.)

UNITED STATES OF AMERICA
DEPARTMENT OF TRANSPORTATION — FEDERAL AVIATION ADMINISTRATION

STANDARD AIRWORTHINESS CERTIFICATE

1 NATIONALITY AND REGISTRATION MARKS	2 MANUFACTURER AND MODEL	3 AIRCRAFT SERIAL NUMBER	4 CATEGORY
N12345	Flitmore FT-3	6969	NORMAL

5 AUTHORITY AND BASIS FOR ISSUANCE

This airworthiness certificate is issued pursuant to the Federal Aviation Act of 1958 and certifies that as of the date of issuance the aircraft to which issued has been inspected and found to conform to the type certificate therefor, to be in condition for safe operation, and has been shown to meet the requirements of the applicable comprehensive and detailed airworthiness code as provided by Annex 8 to the Convention on International Civil Aviation, except as noted herein.

Exceptions:

NONE

6 TERMS AND CONDITIONS

Unless sooner surrendered, suspended, revoked, or a termination date is otherwise established by the Administrator, this airworthiness certificate is effective as long as the maintenance, preventative maintenance, and alterations are performed in accordance with Parts 21, 43 and 91 of the Federal Aviation Regulations, as appropriate, and the aircraft is registered in the United States.

DATE OF ISSUANCE	FAA REPRESENTATIVE	DESIGNATION NUMBER
11/15/85	Philipe Cordoba	AEA-GADO-03

Any alteration, reproduction, or misuse of this certificate may be punishable by a fine not exceeding $1,000 or imprisonment not exceeding 3 years, or both. THIS CERTIFICATE MUST BE DISPLAYED IN THE AIRCRAFT IN ACCORDANCE WITH APPLICABLE FEDERAL AVIATION REGULATIONS.

FAA Form 8100-2 (8-82) GPO 892-804

Figure 7. FAA Form 8100–2, Standard Airworthiness Certificate

UNITED STATES OF AMERICA
DEPARTMENT OF TRANSPORTATION — FEDERAL AVIATION ADMINISTRATION

SPECIAL AIRWORTHINESS CERTIFICATE

A	CATEGORY DESIGNATION		
	PURPOSE		
B	MANU-FACTURER	NAME	
		ADDRESS	
C	FLIGHT	FROM	
		TO	
D	N—		SERIAL NO
	BUILDER		MODEL
E	DATE OF ISSUANCE		EXPIRY
	OPERATING LIMITATIONS DATED		ARE A PART OF THIS CERTIFICATE
	SIGNATURE OF FAA REPRESENTATIVE		DESIGNATION OR OFFICE NO

SAMPLE

Any alteration, reproduction or misuse of this certificate may be punishable by a fine not exceeding $1,000 or imprisonment not exceeding 3 years or both. THIS CERTIFICATE MUST BE DISPLAYED IN THE AIRCRAFT IN ACCORDANCE WITH APPLICABLE FEDERAL AVIATION REGULATIONS

FAA FORM 8130-7 (10/82) SEE REVERSE SIDE

Figure 8. FAA Form 8130–7, Special Airworthiness Certificate (pink in color)

No certificate may be issued unless a completed application form has been received (14 C.F.R. 21)

Form Approved O.M.B. No. 04-R0058

DEPARTMENT OF TRANSPORTATION
FEDERAL AVIATION ADMINISTRATION

APPLICATION FOR AIRWORTHINESS CERTIFICATE

INSTRUCTIONS — Print or type. Do not write in shaded areas; these are for FAA use only. Submit original only to an authorized FAA Representative. If additional space is required, use an attachment. For special flight permits complete Sections II and VI or VII as applicable.

I. AIRCRAFT DESCRIPTION

1. REGISTRATION MARK	2. AIRCRAFT BUILDER'S NAME (Make)	3. AIRCRAFT MODEL DESIGNATION	4. YR. MFG.	FAA CODING
N7316	Big Deal, Inc.	BA-OH	1978	
5. AIRCRAFT SERIAL NO.	6. ENGINE BUILDER'S NAME (Make)	7. ENGINE MODEL DESIGNATION		
00021	Twister Engine Corp.	PHP		
8. NUMBER OF ENGINES	9. PROPELLER BUILDER'S NAME (Make)	10. PROPELLER MODEL DESIGNATION	11. AIRCRAFT IS	
2	Fan	TOE1	NEW / X USED / IMPORT	

II. CERTIFICATION REQUESTED

APPLICATION IS HEREBY MADE FOR: (Check applicable items)

A. 1 X STANDARD AIRWORTHINESS CERTIFICATE (Indicate category) X NORMAL / UTILITY / ACROBATIC / TRANSPORT / GLIDER / BALLOON

B. SPECIAL AIRWORTHINESS CERTIFICATE (Check appropriate items)

- 2 LIMITED
- 5 PROVISIONAL (Indicate class): 1 CLASS I; 2 CLASS II
- 3 RESTRICTED (Indicate operation to be conducted): 1 AGRICULTURE AND PEST CONTROL; 2 AERIAL SURVEYING; 3 AERIAL ADVERTISING; 4 FOREST (Wildlife conservation); 5 PATROLLING; 6 WEATHER CONTROL; 0 OTHER (Specify)
- 4 EXPERIMENTAL (Indicate operation to be conducted): 1 RESEARCH AND DEVELOPMENT; 2 AMATEUR BUILT; 3 EXHIBITION; 4 RACING; 5 CREW TRAINING; 6 MKT. SURVEY; 0 TO SHOW COMPLIANCE WITH FAR
- 8 SPECIAL FLIGHT PERMIT (Indicate operation to be conducted, then complete Section VI or VII as applicable on reverse side): 1 FERRY FLIGHT FOR REPAIRS, ALTERATIONS, MAINTENANCE OR STORAGE; 2 EVACUATE FROM AREA OF IMPENDING DANGER; 3 OPERATION IN EXCESS OF MAXIMUM CERTIFICATED TAKE-OFF WEIGHT; 4 DELIVERING OR EXPORT; 5 PRODUCTION FLIGHT TESTING

C. 6 MULTIPLE AIRWORTHINESS CERTIFICATE (Check ABOVE "Restricted Operation" and "Standard" or "Limited" as applicable)

III. OWNER'S CERTIFICATION

A. REGISTERED OWNER (As shown on certificate of aircraft registration) — IF DEALER, CHECK HERE

NAME	ADDRESS
MARION W. WILLIAMS	1000 Whitehouse Road, Oklahoma City, OK

B. AIRCRAFT CERTIFICATION BASIS (Check applicable blocks and complete items as indicated)

X	AIRCRAFT SPECIFICATION OR TYPE CERTIFICATE DATA SHEET (Give No. and Revision No.): BD 4-7X3	X	AIRWORTHINESS DIRECTIVES (Check if all applicable AD's complied with and give latest AD No.): AD #BD 1-01
X	AIRCRAFT LISTING (Give page number(s)): Page 382		SUPPLEMENTAL TYPE CERTIFICATE (List number of each STC incorporated)

C. AIRCRAFT OPERATION AND MAINTENANCE RECORDS

X	CHECK IF RECORDS IN COMPLIANCE WITH FAR 91.173	TOTAL AIRFRAME HOURS (Enter for used aircraft only): 100	EXPERIMENTAL ONLY (Enter hours flown since last certificate issued or renewed)

D. CERTIFICATION — I hereby certify that I am the owner (or his agent) of the aircraft described above, that the aircraft is registered with the Federal Aviation Administration in accordance with Section 501 of the Federal Aviation Act of 1958, and applicable Federal Aviation Regulations, and that the aircraft has been inspected and is airworthy and eligible for the airworthiness certificate requested.

DATE OF APPLICATION	NAME AND TITLE (Print or type)	SIGNATURE
8/6/85	MARION W. WILLIAMS, Owner	Marion W. Williams

IV. INSPECTION AGENCY VERIFICATION

A. THE AIRCRAFT DESCRIBED ABOVE HAS BEEN INSPECTED AND FOUND AIRWORTHY BY: (Complete this section only if FAR 21.183(d) applies)

- FAR PART 121 OR 127 CERTIFICATE HOLDER (Give Certificate No.)
- CERTIFICATED MECHANIC (Give Certificate No.)
- CERTIFICATED REPAIR STATION (Give Certificate No.): 0035-1
- AIRCRAFT MANUFACTURER (Give name of firm)

DATE	TITLE	SIGNATURE
8/10/85	Supervisor, Maintenance Operations	Charles E. Griffin — CHARLES E. GRIFFIN

V. FAA REPRESENTATIVE CERTIFICATION

(Check ALL applicable blocks in items A and B)

A. I find that the aircraft described in Section I or VII meets requirements for — THE CERTIFICATE REQUESTED; AMENDMENT OR MODIFICATION OF CURRENT AIRWORTHINESS CERTIFICATE

B. Inspection for a special flight permit under section VII was conducted by: I.R. Inspector — FAA INSPECTOR; CERTIFICATE HOLDER UNDER; FAR 65; FAR 121 OR 127; FAR 145

DATE	DISTRICT OFFICE	DESIGNEE'S SIGNATURE AND NO.	FAA INSPECTOR'S SIGNATURE
8/13/85	ASO GADO 9	I.R. INSPECTOR IA456551277	M.E. Tu — M.E. TU, ASO GADO 9

FAA Form 8130–6 (4-80) SUPERSEDES PREVIOUS EDITION

Figure 9. FAA Form 8130–6, Application for Airworthiness Certificate

AIRCRAFT MAINTENANCE

Maintenance means the preservation, inspection, overhaul, and repair of aircraft, including the replacement of parts. A PROPERLY MAINTAINED AIRCRAFT IS A SAFE AIRCRAFT.

The purpose of maintenance is to ensure that the aircraft remains airworthy throughout its operational life.

Although maintenance requirements will vary for different types of aircraft, experience shows that most aircraft will need some type of preventive maintenance every 25 hours or less of flying time, and minor maintenance at least every 100 hours. This is influenced by the kind of operation, climatic conditions, storage facilities, age, and construction of the aircraft. Most manufacturers supply service information which should be used in maintaining your aircraft.

Inspections

FAR Part 91 places primary responsibility on the owner or operator for maintaining an aircraft in an airworthy condition. Certain inspections must be performed on your aircraft, and you must maintain the airworthiness of the aircraft between required inspections by having any defects corrected.

FAR's require the inspection of all civil aircraft at specific intervals to determine the overall condition. The interval depends generally upon the type of operations in which the aircraft is engaged. Some aircraft need to be inspected at least once each 12 calendar months, while inspection is required for others after each 100 hours of operation. In other instances, an aircraft may be inspected in accordance with an inspection system set up to provide for total inspection of the aircraft on the basis of calendar time, time in service, number of system operations, or any combination of these.

To determine the specific inspection requirements and rules for the performance of inspections, refer to the FAR's which prescribe the requirements for various types of operations.

Annual Inspection. Any reciprocating–engine powered or single–engine–turbojet/turbopropeller driven small aircraft (12,500 pounds and under) flown for pleasure is required to be inspected at least annually by an FAA certificated A & P mechanic holding an Inspection Authorization, or an FAA certificated repair station that is appropriately rated, or the manufacturer of the aircraft. The aircraft may not be operated unless the annual inspection has been performed within the preceding 12 calendar months. A period of 12 calendar months extends from any day of a month to the last day of the same month the following year. However, an aircraft with the annual inspection overdue may be operated under a special flight permit issued by the FAA for the purpose of flying the aircraft to a location where the annual inspection can be performed.

100–Hour Inspection. Any reciprocating–engine powered or single–engine–turbojet/turbopropeller driven small aircraft (12,500 pounds and under) used to carry passengers or for flight instruction for hire, must be inspected within each 100 hours of time in service by an FAA certificated A & P mechanic, an FAA certificated repair station that is appropriately rated, or the aircraft manufacturer. An annual inspection is acceptable as a 100–hour inspection, but the reverse is not true.

Other Inspection Programs. The annual and 100–hour inspection requirements do not apply to large (over 12,500 pounds) airplanes, turbojet, or turbopropeller–powered multiengine airplanes, or to airplanes for which the owner or operator complies with the progressive inspection requirements. Details of these requirements may be determined by reference to FAR Parts 43 and 91 and by inquiry at a local FAA General Aviation or Flight Standards District Office.

Preflight Inspection. The FAR's require a pilot to conduct a thorough preflight inspection before every flight to assure that the aircraft is safe for flight.

Preventive Maintenance

The FAR's list approximately two dozen relatively uncomplicated repairs and procedures defined as *preventive maintenance.* Certificated pilots, excluding student pilots, may perform preventive maintenance on any aircraft owned or operated by them that are not used in air carrier service. These preventive maintenance operations are listed in FAR Part 43, Maintenance, Preventive Maintenance, Rebuilding, and Alteration. FAR Part 43 also contains other rules to be followed in the maintenance of aircraft.

Repairs and Alterations

All repairs and alterations are classed as either major or minor. Major repairs or major alterations must be approved for return to service by an appropriately rated certificated repair station, an FAA certificated A & P mechanic holding an Inspection Authorization, or a representative of the Administrator. Minor repairs and minor alterations may be approved for return to service by an FAA certificated A & P mechanic or an appropriately certificated repair station.

AIRCRAFT MAINTENANCE RECORDS

(FAR Section 91.173)

The owner of an aircraft is required to keep aircraft maintenance records which contain a description of the work performed on the aircraft, the date the work was completed, and the signature and FAA certificate number of the person approving the aircraft for return to service. The owner's aircraft record must also contain additional information required by FAR Section 91.173.

Proper management of aircraft operations begins with a good system of maintenance records. A properly completed maintenance record provides the information needed by the owner/operator and maintenance personnel to determine when scheduled inspections and maintenance are to be performed.

A. There must be records of maintenance and alterations and records of the 100–hour, annual, progressive, and other required or approved inspections for each aircraft, including the airframe, each engine, propeller, rotor, and appliance. These records may be discarded when the work is repeated or superseded by other work, or 1 year after the work is performed.

B. There must also be records of:

(1) The total time in service of the airframe, each engine, and each propeller;
(2) The current status of life–limited parts of each airframe, engine, propeller, rotor, and appliance;
(3) The time since the last overhaul of all items installed on the aircraft which are required to be overhauled on a specified time basis;
(4) The identification of the current inspection status of the aircraft including the time since the last inspection required by the inspection program under which the aircraft and its appliances are maintained;
(5) The current status of applicable AD's including, for each, the method of compliance, the AD number, and revision date. If the AD involves recurring action, the time and date when the next action is required; and
(6) A copy of the current major alterations to each airframe, engine, propeller, and appliance.

These records must be retained by the owner/operator and must be transferred with the aircraft if it is sold.

Keep in mind that as a result of repairs or alterations, amendments may be necessary to the weight and balance report, equipment list, flight manual, etc.

Entries into the Aircraft Maintenance Records

A. FAR Section 43.9 entries.

Any person who maintains, rebuilds, or alters an aircraft, airframe, aircraft engine, propeller, or appliance must make an entry containing:

(1) A description of the work, or some reference to data acceptable to the FAA;

(2) The date the work was completed;
(3) The name of the person who performed the work; and
(4) If the work is approved for return to service, the signature and certificate number of the person approving the aircraft for return to service.

B. FAR Section 43.11 entries.

When a mechanic approves or disapproves an aircraft for return to service after an annual, 100–hour, or progressive inspection, an entry must be made including:

(1) Aircraft time in service;
(2) The type of inspection;
(3) The date of the inspection;
(4) The certificate number of the person approving or disapproving the aircraft for return to service; and
(5) A signed and dated listing of discrepancies and unairworthy items.

C. FAR Section 91.169(e) — Airplanes.

Inspection entries for FAR Section 91.169(e), Airplanes — those over 12,500 pounds, turbojet, or turbopropeller–powered multiengine airplanes are made according to FAR Section 43.9 and they must include:

(1) The kind of inspection performed;
(2) A statement by the mechanic that it was performed in accordance with the instructions and the procedures for the kind of inspection program selected by the owner; and
(3) A statement that a signed and dated list of any defects found during the inspection was given to the owner, if the aircraft is not approved for return to service.

D. FAA Form 337, Major Repairs and Major Alterations.

A mechanic who performs a major repair or major alteration must record it on FAA Form 337 and have the work inspected and approved by a mechanic who holds an Inspection Authorization. A signed copy must be given to the owner and another copy sent to the local FAA Flight Standards district office within 48 hours after the aircraft has been approved for return to service. However, when a major repair is done by a certificated repair station, the customer's work order may be used and a release given as outlined in Appendix B of FAR Part 43.

E. FAR Section 91.171 entries.

FAR Section 91.171 requires that every airplane operated in controlled airspace under IFR conditions must have each static pressure system and each altimeter tested and inspected each 24 calendar months. The mechanic must enter into the records:

(1) A description of the work;
(2) The maximum altitude to which the altimeter was tested; and
(3) The date and signature of the person approving the airplane for return to service.

Additional Information on Aircraft Maintenance Records

Additional information relating to aircraft maintenance records may be obtained from FAR Part 39—Airworthiness Directives; FAR Part 43—Maintenance, Preventive Maintenance and Alteration; FAR Part 91—General Operating and Flight Rules; and Advisory Circular 43–9 (latest revision), Maintenance Records: General Aviation Aircraft.

These publications are available for review at your local FAA Flight Standards district office where you can obtain assistance in establishing your aircraft maintenance program and the necessary maintenance records.

SPECIAL FLIGHT PERMITS

(FAR Section 21.197)

A special flight permit is a Special Airworthiness Certificate issued for an aircraft that may not currently meet applicable airworthiness requirements, but is safe for a specific flight. Before the permit is issued, an FAA inspector may personally inspect the aircraft or require it to be inspected by an FAA certificated A & P mechanic or repair station to determine its safety for the intended flight. The inspection must be recorded in the aircraft records.

Special flight permits are issued to allow the aircraft to be flown to a base where repairs, alterations, or maintenance can be performed; for delivering or exporting the aircraft; or, for evacuating an aircraft from an area of impending danger. They may also be issued to allow the operation of an overweight aircraft for flight beyond its normal range over water or land areas where adequate landing facilities or fuel are not available.

Should you have occasion to need a special flight permit, assistance and the necessary forms may be obtained from the local FAA Flight Standards or General Aviation District Office. (See Figure 10, for a sample of the special flight permit application form, reverse side of FAA Form 8130–6, Application for Airworthiness Certificate.)

VI. PRODUCTION FLIGHT TESTING

A. MANUFACTURER

NAME

ADDRESS

B. PRODUCTION BASIS (Check applicable item)

- [ ] PRODUCTION CERTIFICATE (Give production certificate number)
- [ ] TYPE CERTIFICATE ONLY
- [ ] APPROVED PRODUCTION INSPECTION SYSTEM

C. GIVE QUANTITY OF CERTIFICATES REQUIRED FOR OPERATING NEEDS

DATE OF APPLICATION | NAME AND TITLE (Print or type) | SIGNATURE

VII. SPECIAL FLIGHT PERMIT PURPOSES OTHER THAN PRODUCTION FLIGHT TEST

A. DESCRIPTION OF AIRCRAFT

REGISTERED OWNER: MARION W. WILLIAMS	ADDRESS: 1000 Whitehouse Road, OKC, OK 73100
BUILDER (Make): Big Deal, Inc.	MODEL: BA-OH
SERIAL NUMBER: 00021	REGISTRATION MARK: N7316

B. DESCRIPTION OF FLIGHT

FROM: Oklahoma City, Oklahoma	TO: Dublin, Ireland	
VIA: Reykjavik, Iceland	DEPARTURE DATE: Sept. 9, 1985	DURATION: 21 days

C. CREW REQUIRED TO OPERATE THE AIRCRAFT AND ITS EQUIPMENT

X PILOT   X CO PILOT   X NAVIGATOR   OTHER (Specify)

D. THE AIRCRAFT DOES NOT MEET THE APPLICABLE AIRWORTHINESS REQUIREMENTS AS FOLLOWS:

Aircraft has auxiliary fuel tank (40 gallon capacity) installed to extend range for this special flight.

E. THE FOLLOWING RESTRICTIONS ARE CONSIDERED NECESSARY FOR SAFE OPERATION (Use attachment if necessary)

Make takeoff and landing from main fuel tank. Use auxiliary tank only in cruise.

F. CERTIFICATION—I hereby certify that I am the registered owner (or his agent) of the aircraft described above, that the aircraft is registered with the Federal Aviation Administration in accordance with Section 501 of the Federal Aviation Act of 1958, and applicable Federal Aviation Regulations, and that the aircraft has been inspected and is airworthy for the flight described.

DATE	NAME AND TITLE (Print or type)	SIGNATURE
Aug. 15, 1985	MARION W. WILLIAMS, Owner	Marion W. Williams

VIII. AIRWORTHINESS DOCUMENTATION (FAA use only)

A. Operating Limitations and Markings in Compliance with FAR 91.31 as Applicable	G. Statement of Conformity, FAA Form 317 (Attach when required)
B. Current Operating Limitations Attached	H. Foreign Airworthiness Certification for Import Aircraft (Attach when required)
C. Data, Drawings, Photographs, etc. (Attach when required)	I. Previous Airworthiness Certificate Issued in Accordance with
D. Current Weight and Balance Information Available in Aircraft	FAR ______ CAR ______ (Original attached)
E. Major Repair and Alteration, FAA 337 (Attach when required)	J. Current Airworthiness Certificate Issued in Accordance with
F. This Inspection Recorded in Aircraft Records	FAR ______ (Copy attached)

U.S. GOVERNMENT PRINTING OFFICE

Figure 10. FAA Form 8130–6, Application for Airworthiness Certificate, (reverse side) Application for Special Flight Permit

AIRWORTHINESS DIRECTIVES

(FAR Part 39)

A primary safety function of the FAA is to require correction of unsafe conditions found in an aircraft, aircraft engine, propeller, or appliance when such conditions exist or are likely to exist or develop in other products of the same design. The unsafe conditions may exist because of a design defect, maintenance, or other causes. FAR Part 39, Airworthiness Directives, defines the authority and responsibility of the Administrator for requiring the necessary corrective action. AD's are the media used to notify aircraft owners and other interested persons of unsafe conditions and to specify the conditions under which the product may continue to be operated.

AD's may be divided into two categories:

1—Those of an emergency nature requiring immediate compliance prior to further flight, and
2—Those of a less urgent nature requiring compliance within a relatively longer period of time.

AD's are FAR's and must be complied with, unless specific exemption is granted. It is the aircraft owner's or operator's responsibility to assure compliance with all pertinent AD's. This includes those AD's that require recurrent or continuing action. For example, an AD may require a repetitive inspection each 50 hours of operation, meaning the particular inspection shall be accomplished and recorded every 50 hours of time in service.

FAR's require a record to be maintained that shows the current status of applicable AD's, including the method of compliance, the AD number and revision date, and the signature and certificate number of the repair station or mechanic who performed the work. For ready reference, many aircraft owners have a chronological listing of the pertinent AD's in the back of their aircraft and engine records.

The Summary of Airworthiness Directives contains all the valid AD's previously published and biweekly supplements. The summary is divided into two volumes. Volume I includes directives applicable to small aircraft (12,500 pounds, or less, maximum certificated takeoff weight). Volume II includes directives applicable to large aircraft (over 12,500 pounds maximum certificated takeoff weight). Subscription service will consist of the summary and automatic biweekly updates to each summary for a 2-year period. The Summary of Airworthiness Directives, Volume I and Volume II, are sold and distributed for the Superintendent of Documents by the FAA from Oklahoma City and are available in either hard copy or microfiche. Requests for subscription service or pricing information to either of these publications should be sent to:

U.S. Department of Transportation
Federal Aviation Administration
Mike Monroney Aeronautical Center
General Accounting Branch, AAC–23
P.O. Box 25461
Oklahoma City, OK 73125

THE SERVICE DIFFICULTY PROGRAM

The Service Difficulty Program provides for the exchange of service experience with aircraft and aircraft products to aid in the detection of mechanical problems. The incentive for early detection is to get a jump on corrective actions and ultimate solutions, thereby minimizing the impact of equipment failure on safety.

Aircraft owners, pilots, and mechanics are urged to report promptly all service problems, using FAA Form 8010–4, Malfunction or Defect Report. Copies of these forms may be obtained free from any FAA Flight Standards or General Aviation District Office. No postage is required. (See Figure 11, page 38.)

Each problem reported contributes to the improvement of aviation safety through the identification of a potential problem area and the alerting of other persons to it. This focusing of attention on a problem has led to improvements in the design and maintainability of aircraft and aircraft products.

How does reporting a problem help you? By pooling everyone's knowledge about a situation, we can detect mechanical problems early enough to correct them before they might possibly result in accidents/incidents. This should make flying safer, more fun, and certainly less expensive.

Advisory Circular 20–109, Service Difficulty Program (General Aviation), contains additional information on this program. (See page 41 for ordering instructions.)

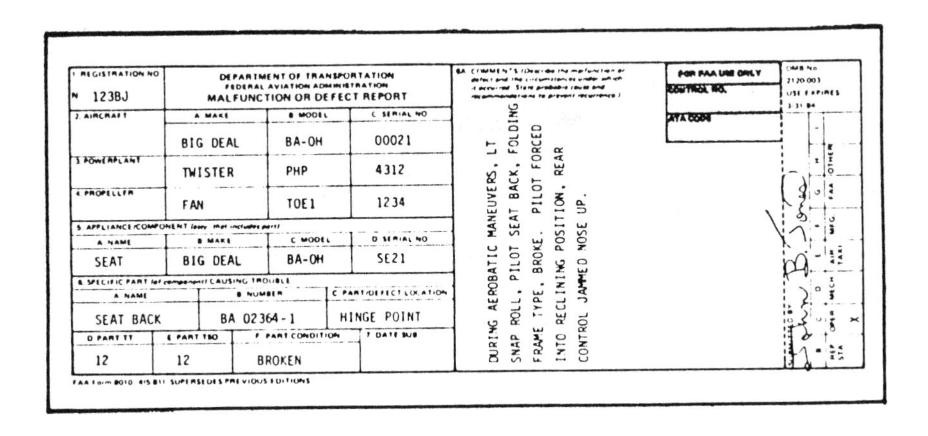

1. REGISTRATION NO: N 123BJ

DEPARTMENT OF TRANSPORTATION
FEDERAL AVIATION ADMINISTRATION
MALFUNCTION OR DEFECT REPORT

	A. MAKE	B. MODEL	C. SERIAL NO.
2. AIRCRAFT	BIG DEAL	BA-OH	00021
3. POWERPLANT	TWISTER	PHP	4312
4. PROPELLER	FAN	TOE1	1234

5. APPLIANCE/COMPONENT

A. NAME	B. MAKE	C. MODEL	D. SERIAL NO.
SEAT	BIG DEAL	BA-OH	SE21

6. SPECIFIC PART (of component) CAUSING TROUBLE

A. NAME	B. NUMBER	C. PART/DEFECT LOCATION
SEAT BACK	BA 02364-1	HINGE POINT

D. PART TT	E. PART TSO	F. PART CONDITION	7. DATE SUB.
12	12	BROKEN	

FAA Form 8010-4 (5-81) SUPERSEDES PREVIOUS EDITIONS

8A. COMMENTS (Describe the malfunction or defect and the circumstances under which it occurred. State probable cause and recommendations to prevent recurrence.)

DURING AEROBATIC MANEUVERS, LT
SNAP ROLL, PILOT SEAT BACK, FOLDING
FRAME TYPE, BROKE. PILOT FORCED
INTO RECLINING POSITION, REAR
CONTROL JAMMED NOSE UP.

FOR FAA USE ONLY
CONTROL NO.
ATA CODE

OMB No. 2120-003
USE EXPIRES 3-31-84

SUBMITTED BY: John B. Jones

B	C	D	E	F	G	H
REP. STA.	OPER.	MECH.	AIR TAXI	MFG.	FAA	OTHER
	X					

Figure 11. FAA Form 8010-4, Malfunction or Defect Report

AC No. 43-16

AIRWORTHINESS ALERTS

The FAA publishes Advisory Circular 43–16, General Aviation Airworthiness Alerts, monthly to provide the aviation community with a means for interchanging service difficulty information. The articles contained in the Alerts are derived from the Malfunction or Defect Reports submitted by aircraft owners, pilots, mechanics, repair stations, and air taxi operators. (See page 41 for instructions on obtaining advisory circulars.)

Maintenance and engineering specialists review the reports and select pertinent items for publication in the Alerts. The information is brief and advisory; compliance is not mandatory. It is, however, intended to alert you to service experience and, when pertinent, direct your attention to the manufacturer's recommended corrective action.

The Alerts are distributed automatically to personnel responsible for approving aircraft for return to service, such as certificated repair stations, mechanics holding an Inspection Authorization, etc. Limited copies of the Alerts may be available at local FAA district offices.

OBTAINING FAA PUBLICATIONS AND RECORDS

Advisory Circulars

Advisory circulars are issued by the FAA to inform the aviation public, in a systematic way, of nonregulatory material of interest. The contents of advisory circulars are not binding on the public unless incorporated into a regulation by reference.

Advisory Circular 00–2, Advisory Circular Checklist, contains a list of current FAA advisory circulars and provides detailed instructions on how to obtain them. It also contains a list of U.S. Government Printing Office Bookstores located throughout the United States which stock many Government publications. This advisory circular may be obtained free upon request from the U.S. Department of Transportation, Subsequent Distribution Section, M–494.3, Washington, DC 20590.

Federal Aviation Regulations

The following regulations are those you may be most interested in reading. They pertain primarily to the operation and maintenance of the aircraft and to obtaining a pilot's certificate or an A & P mechanic certificate.

Part 1	Definitions and Abbreviations
Part 21	Certification Procedures for Products and Parts
Part 23	Airworthiness Standards: Normal, Utility, and Acrobatic Category Aircraft
Part 33	Airworthiness Standards: Aircraft Engines
Part 35	Airworthiness Standards: Propellers
Part 39	Airworthiness Directives
Part 43	Maintenance, Preventive Maintenance, Rebuilding, and Alteration
Part 45	Identification and Registration Marking
Part 47	Aircraft Registration
Part 49	Recording of Aircraft Titles and Security Documents

Part 61	Certification: Pilots and Flight Instructors
Part 65	Certification: Airmen Other Than Flight Crewmembers
Part 91	General Operating and Flight Rules

Advisory Circular 00–44, Status of Federal Aviation Regulations, contains the current status of the FAR's including changes issued, price list, and ordering instructions. This advisory circular may be obtained free upon request from the U.S. Department of Transportation, Subsequent Distribution Section, M–494.3, Washington, DC 20590.

Records

If you become an aircraft owner, pilot, or certificated mechanic, you may, at some time, need to obtain copies of documents pertaining to your aircraft, airman, or medical certification. Aircraft records are available from the FAA Aircraft Registry, AAC–250, (405) 686–2116; airman records from Airmen Certification Branch, AAC–260; and medical records from Aeromedical Certification Branch, AAC–130. Fees for furnishing copies of paper records are: $2 for search, $0.25 for copy of first page, $0.05 for second and each additional page; $0.15 for each microfiche for microfiched records; $3 for certification of copies as duplicates of the original records; $2 for duplicate aircraft registration or airman certification or medical certificate; and $5 for certification of diligent search (search of all possible sources of information).

Fees, which are subject to change, may be paid by check, draft, or postal money order, payable to the Treasurer of the United States. Send your request to the proper branch at the following address:

U.S. Department of Transportation
Federal Aviation Administration
Mike Monroney Aeronautical Center
P.O. Box 25082
(Insert proper branch name and routing symbol.)
Oklahoma City, OK 73125

If a prospective owner has reason to believe that an aircraft has been previously destroyed or demolished and has been rebuilt or restored, the FAA Aircraft Registry, AAC–250, 6500 South MacArthur Boulevard, Oklahoma City, OK 73125, will have documentation if the aircraft was reported to the FAA as destroyed or demolished. The aircraft records may be requested and reviewed at the above address.

FAA DISTRICT OFFICES

FAA GADO's (General Aviation District Offices) and FSDO's (Flight Standards District Offices) are listed by state wherein their area of responsibility is assigned. Any contacts with a district office should be made to the office nearest your residence. If the responsibility for your locality is not in that office, you will be advised which office to contact.

ALABAMA

FSDO 67B
Municipal Airport
FSS/WB Building
6500 43rd Avenue North
Birmingham, AL 35206
Phone: (205) 254-1557

ALASKA

FSDO 61
3788 University Avenue
Fairbanks, AK 99701
Phone: (907) 452-1276

FSDO 62
A.I.R. Center Building
9610 Shell Simmons Drive
Juneau, AK 99803
Phone: (907) 789-0231

FSDO 63
6601 South Airpark Place
Suite 216
Anchorage, AK 99502
Phone: (907) 243-1902

ARIZONA

FSDO 7
Scottsdale Municipal Airport
15041 North Airport Drive
Scottsdale, AZ 85260
Phone: (602) 241-2561

ARKANSAS

FSDO 65
Adams Field
FAA Building, Room 201
Little Rock, AR 72202
Phone: (501) 378-5565

CALIFORNIA

FSDO 1
Van Nuys Airport
Suite 316
7120 Hayvenhurst Avenue
Van Nuys, CA 91406
Phone: (818) 904-6291

FSDO 2
San Jose Municipal Airport
1387 Airport Boulevard
San Jose, CA 95110
Phone: (408) 291-7681

FSDO 4
Fresno Air Terminal
Suite 110
4955 East Anderson
Fresno, CA 93727
Phone: (209) 487-5306

FSDO 5
Long Beach Airport
2815 East Spring Street
Long Beach, CA 90806
Phone: (213) 426-7134

FSDO 8
Riverside Municipal Airport
6961 Flight Road
Riverside, CA 92504
Phone: (714) 351-6701

FSDO 9
Montgomery Field Airport
Suite 110
8665 Gibbs Drive
San Diego, CA 92123
Phone: (619) 293-5281

FSDO 10
5885 West Imperial Highway
Los Angeles, CA 90045
Phone: (213) 215-2150

FSDO 12
Sacramento Executive Airport
6107 Freeport Boulevard
Sacramento, CA 95822
Phone: (916) 551-1721

FSDO 14
Oakland International Airport
Earhart Road, Building L-105
P.O. Box 2397, Airport Station
Oakland, CA 94614
Phone: (415) 273-7155

COLORADO

FSDO 60 (General Aviation)
Jefferson County Airport
FAA Building 1
Broomfield, CO 80020
Phone: (303) 466-7326

FSDO 60
10455 East 25th Avenue
Suite 202
Aurora, CO 80010
Phone: (303) 340-5400

CONNECTICUT

FSDO 63
Barnes Municipal Airport
Administration Building, First Floor
Westfield, MA 01085
Phone: (413) 568–3121

DELAWARE

FSDO 63
Scott Plaza No. 2
Fourth Floor
Philadelphia, PA 19113
Phone: (215) 596–0673

DISTRICT OF COLUMBIA

FSDO 62
Dulles International Airport
600 West Service Road
Chantilly, VA 20041
Phone: (202) 557–5360

FLORIDA

FSDO 64
St. Petersburg–Clearwater Airport
Terminal Building, West Wing
Clearwater, FL 33520
Phone: (813) 531–1434

FSDO 65
Miami International Airport
Perimeter Road and N.W. 20th Street
FAA Building 3050
P.O. Box 592015
Miami, FL 33159
Phone: (305) 526–2607

FSDO 64J
Craig Municipal Airport
FAA Building
855 St. John's Bluff Road
Jacksonville, FL 32211
Phone: (904) 641–7311

GEORGIA

FSDO 67
3420 Norman Berry Drive
Suite 430
College Park, GA 30354
Phone: (404) 763–7265

HAWAII

FSDO 13
Honolulu International Airport
Air Service Corporation Building
Room 215
218 Lagoon Drive
Honolulu, HI 96819
Phone: (808) 836–0615

IDAHO

FSDO 67A
Boise Airport
3975 Rickenbacker Street
Boise, ID 83705
Phone: (208) 334–1238

ILLINOIS

GADO 3
DuPage County Airport
P.O. Box H
West Chicago, IL 60185
Phone: (312) 377–4516

GADO 19
Capitol Airport
No. 3 North Airport Drive
North Quadrant
Springfield, IL 62708
Phone: (217) 492–4238

INDIANA

GADO 10
Indianapolis International Airport
6801 Pierson Drive
Indianapolis, IN 46241
Phone: (317) 247–2491

GADO 18
Michiana Regional Airport
1843 Commerce Drive
South Bend, IN 46628
Phone: (219) 236–8480

IOWA

FSDO 61
3021 Army Post Road
Des Moines, IA 50321
Phone: (515) 285–9895

KANSAS

FSDO 64
Mid Continent Airport
Flight Standards Building
Wichita, KS 67209
Phone: (316) 926-4462

KENTUCKY

FSDO 63L
Bowman Field
FAA Building
Louisville, KY 40205
Phone: (502) 582-6116

LOUISIANA

FSDO 62
Ryan Airport
9191 Plank Road
Baton Rouge, LA 70811
Phone: (504) 356-5701

MAINE

FSDO 65
Portland International Jetport
General Aviation Terminal
Portland, ME 04102
Phone: (207) 774-4484

MARYLAND

GADO 21
Baltimore-Washington
International Airport
North Administration Building
Elm Road
Baltimore, MD 21240
Phone: (301) 859-5780

MASSACHUSETTS

FSDO 61
Civil Air Terminal Building
Second Floor
Hanscome Field
Bedford, MA 01730
Phone: (617) 273-7231

FSDO 63
Barnes Municipal Airport
Administration Building
Westfield, MA 01085
Phone: (413) 568-3121

MICHIGAN

GADO 8
Kent County International Airport
5500 44th Street, S.E.
Grand Rapids, MI 49508
Phone: (616) 456-2427

FSDO 63
Willow Run Airport
8800 Beck Road
Belleville, MI 48111
Phone: (313) 485-2550

MINNESOTA

GADO 14
Minneapolis-St. Paul International Airport
Room 201
6201 34th Avenue South
Minneapolis, MN 55450
Phone: (612) 725-3341

MISSISSIPPI

FSDO 63J
Jackson Municipal Airport
FAA Building
P.O. Box 6273, Pearl Branch
Jackson, MS 39208
Phone: (601) 960-4633

MISSOURI

FSDO 62
FAA Building
9275 Genaire Drive
Berkeley, MO 63134
Phone: (314) 425-7102

FSDO 63
Kansas City International Airport
525 Mexico City Avenue
Kansas City, MO 64153
Phone: (816) 243-3800

MONTANA

FSDO 65
Helena Airport
FAA Building, Room 3
Helena, MT 59601
Phone: (406) 449-5270

FSDO 65A
Billings Logan International Airport
Administration Building, Room 216
Billings, MT 59101
Phone: (406) 245-6179

NEBRASKA

FSDO 65
Lincoln Municipal Airport
General Aviation Building
Lincoln, NE 68524
Phone: (402) 471-5485

NEVADA

FSDO 6
241 East Reno Avenue
Suite 200
Las Vegas, NV 89119
Phone: (702) 388-6482

FSDO 11
601 South Rock Boulevard
Suite 102
Reno, NV 89502
Phone: (702) 784-5321

NEW HAMPSHIRE

FSDO 65
Portland International Jetport
General Aviation Terminal
Portland, ME 04102
Phone: (207) 774-4484

NEW JERSEY

FSDO 61
Teterboro Airport
150 Riser Road
Teterboro, NJ 07608
Phone: (201) 288-1745

NEW MEXICO

FSDO 61
2402 Kirtland Drive, S.E.
Albuquerque, NM 87106
Phone: (505) 247-0156

NEW YORK

GADO 1
Albany County Airport
CFR & M Building
Albany, NY 12211
Phone: (518) 869-8482

GADO 11
Republic Airport
Administration Building
Farmingdale, NY 11735
Phone: (516) 694-5530

GADO 17
Rochester Monroe County Airport
1295 Scottsville Road
Rochester, NY 14624
Phone: (716) 263-5880

NORTH CAROLINA

FSDO 66
Smith Reynolds Airport
Terminal Building
Second Floor
Winston Salem, NC 27105
Phone: (919) 761-3147

GADO 66C
Douglas Municipal Airport
FAA Building
5318 Morris Field Drive
Charlotte, NC 28208
Phone: (704) 392-3214

FSDO 66R
Raleigh-Durham Terminal B
Route 1, Box 486A
Morrisville, NC 27560
Phone: (919) 755-4240

NORTH DAKOTA

FSDO 64
Hector Airport
Administration Building, Room 216
P.O. Box 5496
Fargo, ND 58105
Phone: (701) 232-8949

OHIO

FSDO 65
4242 Airport Road
Lunken Airport Executive Building
Cincinnati, OH 45226
Phone: (513) 684-2183

FSDO 65
Cleveland Hopkins International Airport
Federal Facilities Office Building
Cleveland, OH 44135
Phone: (216) 267-0220

FSDO 65
Port Columbus International Airport
Lane Aviation Building, Room 234
4393 East 17 Avenue
Columbus, OH 43219
Phone: (614) 469-7476

OKLAHOMA

FSDO 67
Wiley Post Airport
FAA Building, Room 111
Bethany, OK 73008
Phone: (405) 789-5220

OREGON

FSDO 64
Portland-Hillsboro Airport
3355 N.E. Cornell Road
Hillsboro, OR 97124
Phone: (503) 221-2104

FSDO 64A
Mahlon-Sweet Airport
90606 Greenhill Road
Eugene, OR 97402
Phone: (503) 688-9721

PENNSYLVANIA

GADO 3
Allentown-Bethlehem-Easton Airport
RAS Aviation Center Building
Allentown, PA 18103
Phone: (215) 264-2888

GADO 10
Capitol City Airport
Administration Building
Room 201
New Cumberland, PA 17070
Phone: (717) 782-4528

GADO 14
Allegheny County Airport
Administration Building
Room 213
West Mifflin, PA 15122
Phone: (412) 462-5507

FSDO 63
Scott Plaza No. 2
Fourth Floor
Philadelphia, PA 19113
Phone: (215) 596-0673

PUERTO RICO

FSDO 61
Puerto Rico International Airport
Room 203A
San Juan, PR 00913
Phone: (809) 791-5050

RHODE ISLAND

FSDO 63
Barnes Municipal Airport
Administration Building
Westfield, MA 01085
Phone: (413) 568-3121

SOUTH CAROLINA

FSDO 67C
Columbia Metropolitan Airport
2819 Aviation Way
West Columbia, SC 29169
Phone: (803) 765-5931

SOUTH DAKOTA

FSDO 66
Rapid City Regional Airport
Rural Route 2, Box 4750
Rapid City, SD 57701
Phone: (605) 343-2403

TENNESSEE

FSDO 63
International Airport
General Aviation Building, Room 137
2488 Winchester Road
Memphis, TN 38116
Phone: (901) 521-3820

FSDO 63N
Nashville Metropolitan Airport
Room 101
322 Knapp Boulevard
Nashville, TN 37217
Phone: (615) 251-5661

TEXAS

FSDO 63
Love Field Airport
8032 Aviation Place
Dallas, TX 75235
Phone: (214) 357-0142

FSDO 64
Hobby Airport
Room 152
8800 Paul B. Koonce Drive
Houston, TX 77061
Phone: (713) 643-6504

FSDO 66
International Airport
Route 3, Box 51
Lubbock, TX 79401
Phone: (806) 762-0335

FSDO 68
International Airport
Room 201
1115 Paul Wilkins Road
San Antonio, TX 78216
Phone: (512) 824-9535

FSDO 68SA
Miller International Airport
Terminal Building
2600 South Main Street
McAllen, TX 78503
Phone: (512) 682-4812

UTAH

FSDO 67
116 North 2400 West
Salt Lake City, UT 84116
Phone: (801) 524-4247

VERMONT

FSDO 65
Portland International Jetport
General Aviation Terminal
Portland, ME 04102
Phone: (207) 774-4484

VIRGINIA

GADO 16
Byrd International Airport
Terminal Building, Second Floor
Sandston, VA 23150
Phone: (804) 222-7494

VIRGIN ISLANDS

FSDO 61
Puerto Rico International Airport
Room 203A
San Juan, PR 00913
Phone: (809) 791-5050

WASHINGTON

FSDO 61
7300 Perimeter Road South
Seattle, WA 98108
Phone: (206) 431-2742

FSDO 61A
5620 East Rutter Avenue
Spokane, WA 99206
Phone: (509) 456-4618

WEST VIRGINIA

GADO 22
Kanawha Airport
301 Eagle Mountain Road
Charleston, WV 25311
Phone: (304) 343-4689

WISCONSIN

FSDO 61
General Mitchell Field
FAA/WB Building
5300 South Howell Avenue
Milwaukee, WI 53207
Phone: (414) 747-5531

WYOMING

FSDO 60A
Natrona County International Airport
FAA/WB Building
Casper, WY 82601
Phone: (307) 234-8959

☆ U. S. Government Printing Office: 1987 - 202-885 (74751)

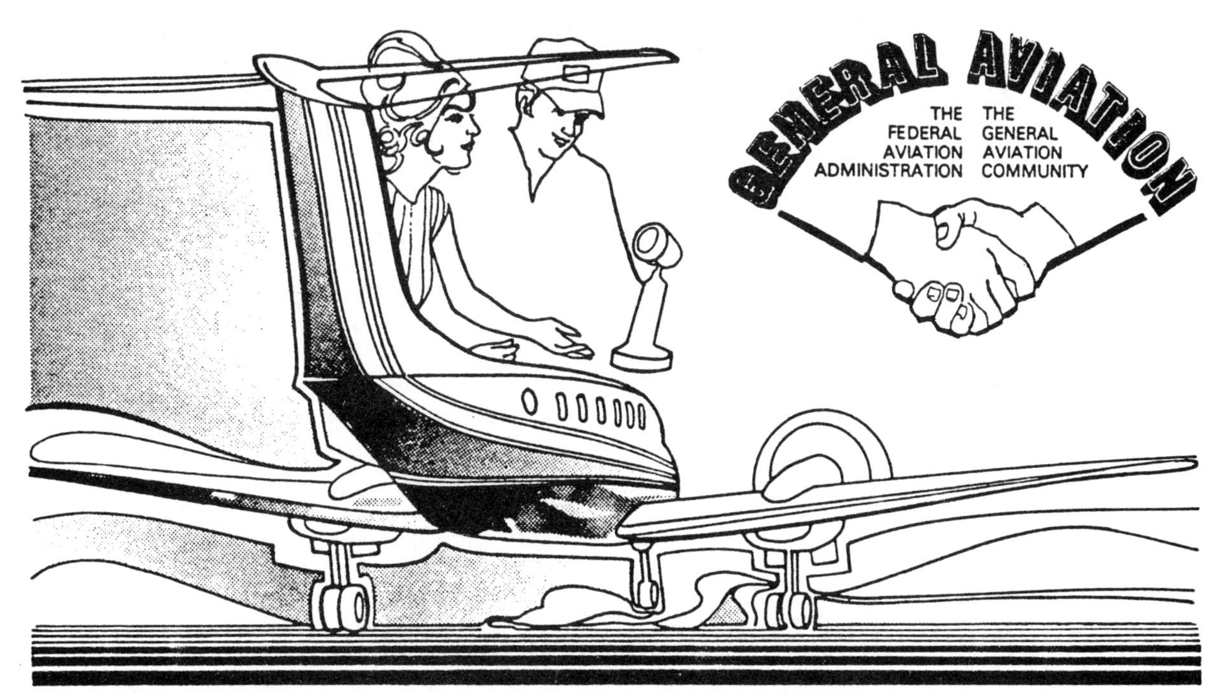

accident prevention program

Maintenance Aspects of Owning Your Own Airplane

U.S. Department of Transportation
Federal Aviation Administration
Washington D.C.

FOREWORD

The purpose of this series of publications is to provide the flying public with safety information that is handy and easy to review. Many of the publications in this series summarize material associated with the audiovisual presentations used in General Aviation Accident Prevention Program activities. Many of these audiovisual presentations were developed through a cooperative project of the FAA, the General Aviation Manufacturers Association, and Association member companies.

Comments regarding these publications should be directed to the Department of Transportation, Federal Aviation Administration, General Aviation Division, Accident Prevention Staff, AFO-806, 800 Independence Avenue, S.W., Washington, D.C. 20591.

INTRODUCTION

As an owner-pilot, FAR Part 43 allows you to perform certain types of inspections and maintenance on your airplane. Here is a partial list of what you can do. See Appendix A of FAR Part 43 for a more complete list.

1. Repair or change tires and tubes.
2. Clean, grease, or replace landing gear wheel bearings.
3. Add air or oil to landing gear shock struts.
4. Replace defective safety wire and cotter keys.
5. Lubricate items not requiring disassembly (other than removal of nonstructural items such as cover plates, cowling, or fairings).
6. Replenish hydraulic fluid.
7. Refinish the exterior or interior of the aircraft (excluding balanced control surfaces) when removal or disassembly of any primary structure or operating system is not required.
8. Replace side windows and safety belts.
9. Replace seats or seat parts with approved replacement parts.
10. Replace bulbs, reflectors, and lenses of position and landing lights.
11. Replace cowling if removal of the propeller is not required.
12. Replace, clean, or set spark plug clearances.
13. Replace hose connections, except hydraulic connections.
14. Replace prefabricated fuel lines.
15. Replace the battery and check fluid level and specific gravity.

Although the above work is allowed by FAR, each individual should make a self analysis as to whether or not he has the ability to perform the work satisfactorily.

If any of the above work is accomplished, an entry must be made in the appropriate logbook. The entry shall contain:

1. A description of the work performed (or references to data that is acceptable to the Administrator).
2. Date of completion.
3. Name of the person performing the work.
4. Signature, certificate number, and kind of certificate held by the person performing the work.

The signature constitutes approval for return to service ONLY for work performed.

INSPECTION CHECK LIST

As a pilot, you may use the following checklist to conduct an inspection of a typical general aviation airplane. Additional copies can be obtained from your FAA General Aviation District Office (GADO).

Propeller; Inspect:

1. Spinner and back plate for cracks or looseness.
2. Blades for nicks or cracks.
3. Hub for grease or oil leaks.
4. Bolts for security and "safetying."

Engine:

1. Preflight engine.
2. Run-up engine to warm-up and check:
   a. Magnetos for RPM drop and ground-out.
   b. Mixture and throttle controls for operation and ease of movement.
   c. Propeller control for operation and ease of movement.
   d. Engine idle for proper RPM.
   e. Carburetor heat or alternate air.
   f. Alternator output under a load (landing light, etc., in the "on" position).
   g. Vacuum system (if installed) for output.
   h. Temperatures (CHT, Oil, etc.) within proper operating range.
   i. Engine and electric fuel pumps for fuel flow or fuel pressure.
   j. Fuel selector, in all positions, for free and proper operation.
3. Remove engine cowling. Clean and inspect for cracks, loose fasteners, or damage.
4. Check engine oil for quantity and condition. Have oil and oil filter changed at 50-hour intervals by an FAA certificated mechanic.
5. Inspect oil temperature "sensing" unit for leaks, security, and broken wires.
6. Inspect oil lines and fittings for condition, leaks and security, and evidence of chafing.
7. Inspect oil cooler for condition (damage, dirt and air blockage), security leaks, and winterization plate (if applicable).

8. Clean engine.
9. Remove, clean, and inspect spark plugs for wear. Regap and reinstall plugs, moving "top to bottom," and "bottom to top" of cylinders. Be sure to gap and torque plugs to manufacturer's specifications.
10. Inspect magnetos for security, cracks, and broken wires or insulation.
11. Inspect ignition harness for chafing, cracked insulation, and cleanliness.
12. Check cylinders for loose or missing nuts and screws, cracks around cylinder hold-down studs, and for broken cooling fins.
13. Check rocker box covers for evidence of oil leaks and loose nuts or screws.
14. Remove air filter and tap gently to remove dirt particles.
15. Replace air filter.
16. Inspect all air-inlet ducts for condition (no air leaks, holes, etc.)
17. Inspect intake seals for leaks (fuel stains) and clamps for security.
18. Check condition of priming lines and fittings for leaks (fuel stains) and clamps for security.
19. Inspect condition of exhaust stacks, connections, clamps, gaskets, muffler, and heat box for cracks, security, condition, and leaks.
20. Inspect condition of fuel lines for leaks (fuel stains) and security.
21. Drain at least one pint of fuel into a transparent container from the fuel filter and from the fuel tank sump to check for water or dirt contamination.
22. Visually inspect vacuum pump and lines for missing nuts, cracked pump flanges, and security.
23. Inspect crankcase breather tubes and clamps for obstructions and security.
24. Inspect crankcase for cracks, leaks, and missing nuts.
25. Inspect engine mounts for cracks or loose mountings.
26. Inspect engine baffles for cracks, security, and foreign objects.
27. Inspect wiring for security, looseness, broken wires, and condition of insulation.
28. Inspect firewall and firewall seals.
29. Inspect generator or alternator belt for proper tension and fraying.
30. Inspect generator (or alternator) and starter for security and safety of nuts and bolts.
31. Inspect brake fluid for level and proper type.
32. Lubricate engine controls: Propeller, mixture, throttle.
33. Inspect alternate air source "door" or carburetor heat to ensure when "door" is closed it has a good seal. Check "door" operation.
34. Reinstall engine cowling.

Cabin; Inspect:

1. Cabin door, latch and hinges for operation and worn door seals.
2. Upholstery for tears.
3. Seats, seat belts, and adjustment hardware.
4. Trim operation for function and ease of movement.
5. Rudder pedals and toe brakes for operation and security.
6. Parking brake.
7. Control wheels, column, pulleys and cables for security, operation and ease of movement.
8. Lights for operation.
9. Heater and defroster controls for operation and ducts for condition and security.
10. Air vents for general condition and operation.
11. Plexiglass in windshield, doors, and side windows for cracks, leaks, and crazing.
12. Instruments and lines for proper operation and security.

Fuselage and Empennage; Inspect:

1. Baggage door, latch, and hinges for security and operation, baggage door seal for wear.
2. Battery for water, corrosion, and security of cables.
3. Antenna mounts and electric wiring for security and corrosion.
4. Hydraulic system for leaks, security, and fluid level.
5. ELT for security, switch position, and battery condition and age.
6. Rotating beacon for security and operation.
7. Stabilizer and control surfaces, hinges, linkages, trim tabs, cables and balance weights for condition, cracks, frayed cables, loose rivets, etc.
8. Control hinges for appropriate lubrication.
9. Static parts for obstructions.

Wings; Inspect:

1. Wing tips for cracks, loose rivets and security.
2. Position lights for operation.
3. Aileron and flap hinges and actuators for cleanliness and lubrication.

4. Aileron balance weights for cracks and security.
5. Fuel tanks, caps and vents, and placards for quantity and type of fuel.
6. Pitot or pitot-static for security and obstruction.

Landing Gear; Inspect:

1. Strut extension.
2. Scissors and nose gear shimmy damper for leaks and loose or missing bolts.
3. Wheels and tires for cracks, cuts, wear and pressure.
4. Hydraulic lines for leaks and security.
5. Gear structure for cracks, loose or missing bolts, and security.
5. Retracting mechanism and gear door for loose or missing bolts and for abnormal wear.
7. Brakes for wear, security, and hydraulic leaks.

Functional Check Flight (FCF); Check:

1. Brakes for proper operation during taxi.
2. Engine and propeller for power, smoothness, etc.; during run-up.
3. Engine instruments for proper reading.
4. Power output (on takeoff run).
5. Flight instruments.
6. Gear retraction and extension for proper operation and warning system.
7. Electrical system (lights; alternator output).
8. Flap operation.
9. Trim functions.
10. Avionics equipment for proper operation (including a VOR or VOT check for all VOR receivers).
11. Operation of heater, defroster, ventilation and air conditioner.

GENERAL

1. Ensure that all applicable A.D.'s have been met and properly recorded in the aircraft records.
2. Comply with applicable service bulletins and service letters.
3. See that the FAA approved *Flight Manual* or *Pilot's Operating Handbook* is aboard and that all required placards are properly installed.
4. See that the Certificate of Airworthiness and aircraft registration are displayed and that the FCC license is aboard.
5. Verify that all FAA required tests involving the transponder, the VOR, and static system have been made and entered in the appropriate aircraft records.

SUMMARY

- It pays to take good care of your engine. Good maintenance is not cheap, but poor performance can be disastrously expensive.
- If you are unqualified or unable to do a particular needed job, depend on competent and certificated mechanics and use approved parts.
- You can save money and have better understanding of your airplane if you participate in the maintenance yourself.
- If you do some of your own maintenance, do it properly. Make sure you complete the job you started.
- Money, time and effort spent on maintenance pays off with your airplane having a higher resale value if you decide to sell.
- Remember, a well cared for airplane is a safe airplane if flown by a competent and proficient pilot. Maintain both your airplane and yourself in top-notch condition.

☆U.S. Government Printing Office: 1984—461-816/10143

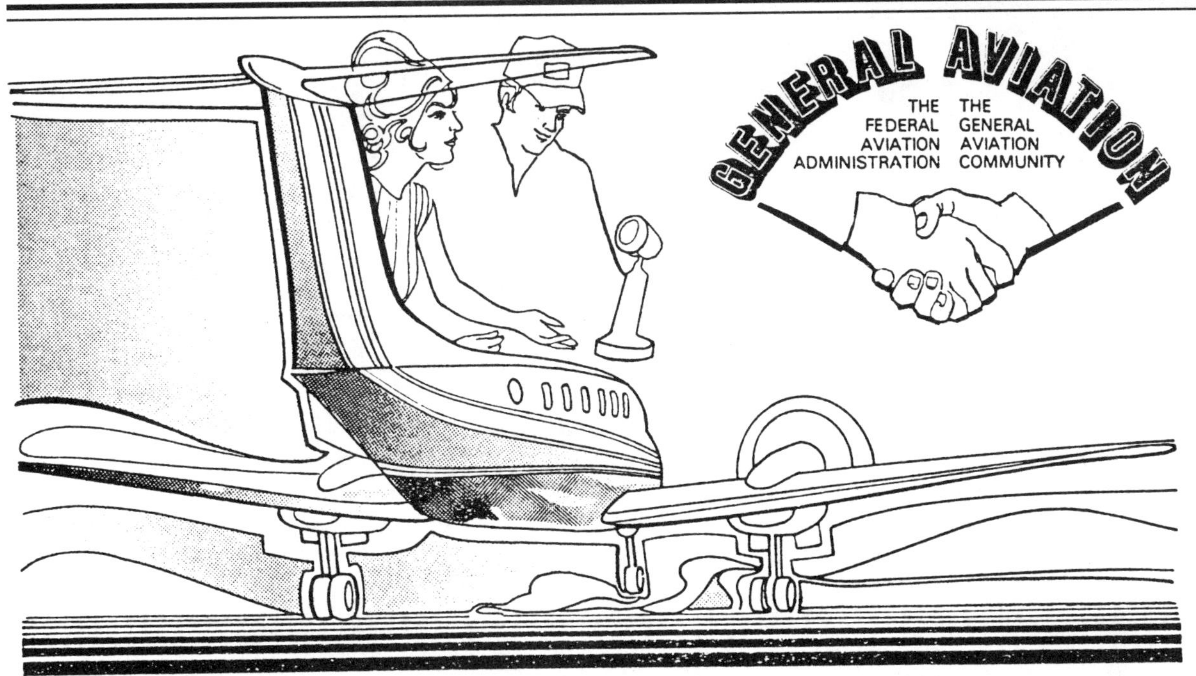

accident prevention program

Preflighting Your Avionics

U.S. DEPARTMENT OF TRANSPORTATION
FEDERAL AVIATION ADMINISTRATION
Washington, D.C.

FOREWORD

The purpose of this series of publications is to provide the flying public with safety information that is handy and easy to review. Many of the publications in this series summarize material associated with the audiovisual presentations used in General Aviation Accident Prevention Program activities. Many of these audiovisual presentations were developed through a cooperative project of the FAA, the General Aviation Manufacturers Association, and Association member companies.

Comments regarding these publications should be directed to the Department of Transportation, Federal Aviation Administration, General Aviation Division, Accident Prevention Staff, AFS–806, 800 Independence Avenue, S.W., Washington, D.C. 20591.

A Cooperative Project by the:

Federal Aviation Administration
General Aviation Manufacturers Association
NARCO Avionics

This presentation is one in a series of industry sponsored safety programs in support of FAA's Accident Prevention Program. Have you seen:

- ☐ "Descent to the MDA and Beyond"
- ☐ "Don't Flirt—Skirt'em"
- ☐ "Engine Operation for Pilots"
- ☐ "Facts of Twin Engine Flying"
- ☐ "General Aviation Normally Aspirated, Direct Drive, Engine Operation"
- ☐ "Handle Like Eggs"
- ☐ "How to Fly Your HSI"
- ☐ "Maintenance Aspects of Owning Your Own Airplane"
- ☐ "Multi-Engine Emergency Procedures"
- ☐ "Pilot Prerogatives"
- ☐ "Preflighting Your Avionics"
- ☐ "Propeller Operation and Care"
- ☐ "Stepping Up to a Complex Airplane"
- ☐ "Take Off Performance Considerations for the Single Engine Airplane"
- ☐ "Time in Your Tanks"
- ☐ "Weatherwise: Go or No Go?"
- ☐ "Why V_{SSE}?"

AVIONICS—is a term used to describe electronics equipment in aircraft.

It includes radios, instruments, and flight control equipment (i.e., autopilots), and *all* of the components required to make up each individual system.

REMEMBER:

- Maximum operating capability and safety is assured *only* when all avionics units are properly:
  —Manufactured,
  —Installed,
  —Maintained, and
  —Operated.
- So, during your initial checkout, familiarize yourself with *all* of the systems in your particular airplane—including the avionics.

- Antennas come with different shapes and locations—so know *your* aircraft.

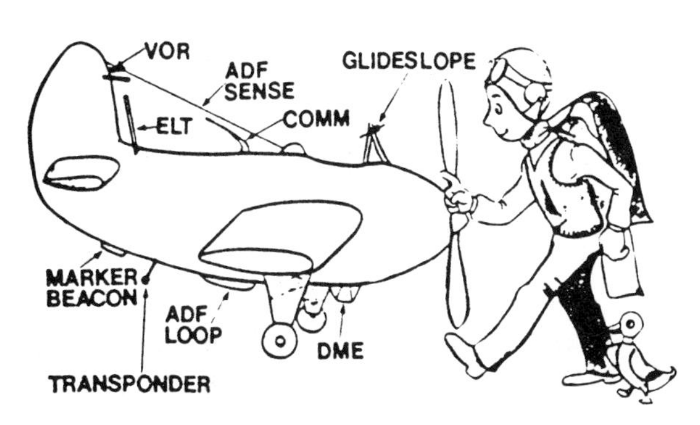

Figure 1. Preflighting your avionics—some typical antenna locations

CHECKING YOUR AVIONICS:

VOR Accuracy

Checking VORs for IFR flight is required under FAR 91.25—within the preceding *thirty* days before flight. Permissible indicated bearing error is as follows:

- With a VOT (VOR test facility), read zero degrees "from" or 180 degrees "to"; maximum error allowed is ±4 degrees.
- Using a VOR check point (see the *Airport Facility Directory* for locations):
  —On the ground, set-in and center the appropriate radial. Maximum permissible error is ±4 degrees.
  —While airborne and over a recognizable landmark, the VOR should read within ±6 degrees.
- If the aircraft is equipped with dual VOR receivers, each VOR display should read within ±4 degrees of the other, when set to the same radial.

DME Accuracy

- Suggested tolerances from the *Airman's Information Manual;* 3% or ½ mile, whichever is greater.

VOR and DME Idents

- DME—every 37½ seconds (higher pitch)
- VOR—4 *code* idents, then DME ident
  or—3 *voice* idents, then DME ident.

SOME GENERAL INFORMATION ABOUT VORs:

- VOR course sensitivity is about 10 degrees from center to full scale deflection.
- ILS localizer course-width sensitivity is about 2½ degrees from center to full scale deflection.

PRELIGHTING YOUR AVIONICS: Walk-around

1. Check the following antennas for physical condition, cracks, oil or dirt, proper mounting, and damage (see figure 1):
   a. Comm or comm nav
   b. VOR
   c. Transponder
   d. Marker beacon
   e. Glide slope
   f. ADF
   g. ELT

In Aircraft, Have On Board:

1. Airworthiness Certificate
2. Registration Certificate
3. FCC Station License
4. *Pilot's Operating Handbook* or *Flight Manual* (operating limitations)
5. Navigation charts and equipment

Before Starting Engine, Check:

1. Avionics equipment-off
2. Pilot heat
3. Magnetic compass, correction card and fluid level
4. Altimeter—set for field elevation, note error
5. Vertical Speed Indicator—on zero
6. Clock—set time

After Starting Engine, Check:

1. All avionics—on
2. Vacuum (suction)—within limits

3. Gyro instruments for:
   a. Erection
   b. Noise
   c. Precession
4. Heading indicator—set
5. Communication radios
   a. Proper frequencies—set
   b. Audio switches—select either speaker or phone
   c. Squelch control—adjust
   d. Volume—adjust
   e. Transmitter select—on desired transmitter
   f. Listen—before transmitting
   g. Microphone—hold close to mouth
6. VOR radios
   a. Proper frequency—set
   b. Flags
   c. Identification
   d. Accuracy, when possible
7. DME
   a. Readout, when possible
   b. Identification
8. ILS
   a. Frequency—set
   b. Flags—localizer and glide slope
   c. Identification
9. Marker beacon
   a. Lights—test, then set on high or low sensitivity.
   b. Audio—on, then adjust volume, when possible.
10. Transponder
    a. Code—set
    b. Switch—set standby
    c. Circuitry—if test switch is provided
11. ADF
    a. Frequency—set
    b. Identification—check
    c. Select ADF mode—then confirm accuracy, when possible.
12. ELT
    a. Set comm radio to 121.5 mHz—listen, check for inadvertent actuation.

While Taxiing, Check:

1. Turn coordinator (or turn indicator)
2. Heading indicator
3. Attitude indicator
4. Autopilot

Before Takeoff:

1. Transponder—on, just before takeoff.

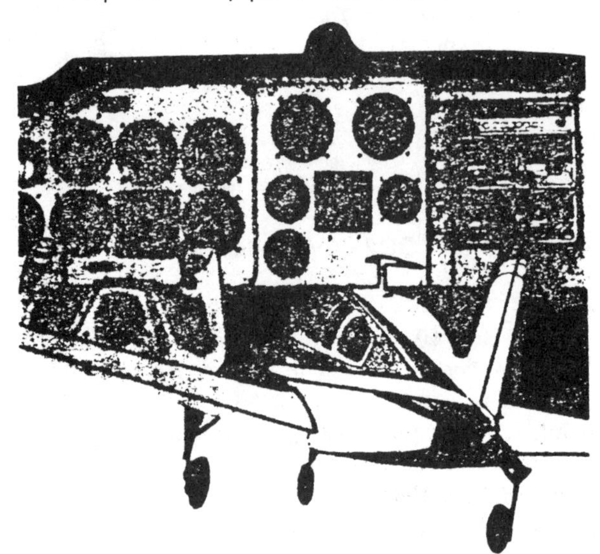

Avco Lycoming TEXTRON
Williamsport Division
Avco Lycoming/Subsidiary of Textron Inc.
652 Oliver Street
Williamsport, PA 17701 U.S.A.
717/323-6181

SERVICE LETTER

Service Letter No. L201B
(Supersedes Service Letter No. L201A)
October 31, 1986

SUBJECT: Recommended Time Between Overhaul Periods.

MODELS AFFECTED: All Avco Lycoming Textron Piston Engines.

The following chart shows the factory-recommended time between overhaul (TBO) for Avco Lycoming Textron piston aircraft engines. The TBO's are based on the installation of GENUINE AVCO LYCOMING TEXTRON PARTS ONLY, average experience in operation, continuous service, and economic factors at the time of engine overhaul. Because of variations in the manner in which engines are operated and maintained, Avco Lycoming Textron can give no assurance that any individual operator will achieve the recommended TBO.

Continuous service assumes that the aircraft will not be out of service for any extended period of time. Refer to latest Service Letter No. L180 if the aircraft is to be out of service for any period of time greater than 30 days.

Although the TBO's shown in the Table on page 2 represent Avco Lycoming Textron's recommendations, the party responsible for maintaining the engine may continue beyond the hours stated, unless otherwise limited by FAA regulations. Furthermore, it is the obligation of the party responsible for maintaining the engine to decide if it should be operated beyond the recommended number of hours.

This decision should be based on knowledge of the engine and the conditions under which it has been operated.

Engine accessories and propellers may require overhaul prior to engine overhaul. This decision, too, is to be made by the party responsible for maintaining the engine or by the accessory manufacturer.

The TBO's in the chart do not apply to engines engaged in crop dusting or other chemical-application flying. These engines should be overhauled at 1500-hour intervals or at recommended TBO, whichever is lower.

Reliability and average service life cannot be predicted when an engine has undergone any modification not approved by Avco Lycoming Textron. The TBO's shown in the table are recommendations for engines as manufactured, without considering any modifications that may alter the life of the engine. The TBO's in no way affect, change, or alter Avco Lycoming Textron's warranty policy or prorated engine replacement policy. See the latest edition of Service Letter No. L191.

RECOMMENDED TIME BETWEEN OVERHAUL PERIODS

FIXED WING AIRCRAFT

Engine Models	See Note	Hours
O-235 Series (except -F, -G, -J)	12	2400
O-235-F, -G, -J	13	2000
O-290-D	-----	2000
O-290-D2	-----	1500
O-320-H	10	2000
O-320; IO-320-A, -E	1,10	2000
IO-320-B, -D, -F	4,6,10	2000
IO-320-C	2,4,10	1800
AIO-320 (160 hp)	6,10	1600
AEIO-320 Series	6,10	1600
O-340 Series	1	2000
O-360; IO-360-B, -E, -F (180 hp)	1,4,10	2000
O-360-E, LO-360-E	4,10	2000
IO-360-A, -C, -D, -J (200 hp)	5,10	1800
TO-360-C, -F; TIO-360-C	3	1600
TO, LTO-360-E (180 hp)	3,4	1800
AIO-360 (200 hp)	6,10	1200
TIO-360	3	1200
AEIO-360 Series (180 hp)	6,10	1400
AEIO-360 Series (200 hp)	6,10	1200
O-435; GO-435	-----	1200
GO, GSO-480; IGSO-480	1	1400
O-540-A, -B, -E4A5, -F; IO-540-C, -D	1,10	2000
O-540-E4B5, -E4C5	1,10,11	2000
O-540-G, -H, -J; IO-540-N, -T, -V, -W	10	2000
O-540-L3C5D	2,10	2000
IO-540-A, -B (290 hp)	10	1200
IO-540-E, -G, -P	1,10	1400
IO-540-S, -AA1A5	2,10	1800
IO-540-J, -R	2,10	1800
IO-540-K (except -K1B5), -L -M	10	2000
IO-540-K1B5	10,11	2000
AEIO-540 Series	6,10	1400
IGO & IGSO-540 Series	-----	1200
TIO-540-V, -W	3,4,11	2000
TIO-540-A, -C, -S, -U	3,4,7,11	1800
TIO-540-J	3,4,11	1600
TIO-540-F, -N, -R	3,4,11	1600
TIO-541-A (310 hp)	3	1300
TIO-541-E (380 hp)	3,9	1600
TIGO-541 (425 hp)	3	1200
IO-720 Series	11	1800

ROTARY WING AIRCRAFT

Engine Models	See Note	Hours
O-320-A2C, -B2C	-----	2000
O-360-C2B, -C2D; HO-360; HIO-360-B	1	1200
HIO-360-A, -C, -D, -E, -F Series	-----	1500
VO-360-A Series	-----	600
VO-360-B; IVO-360	-----	1000
VO-435-A Series	-----	1200
VO-435-B Series	-----	1200
TVO-435 Series	3	1000
VO-540 Series	8	1200
IVO-540 Series	-----	600
TVO, TIVO-540 Series	3,8	1200

NOTES

1. Only engines built with 1/2 inch dia. exhaust valve stems. Engines of this series with 7/16 inch dia. exhaust valves should not exceed 1200 hours between overhauls. New and remanufactured engines built with 1/2 inch dia. exhaust valve stems are identified, respectively, by serial number and date in the latest edition of Service Instruction No. 1136.

2. These engines are designed to incorporate exhaust turbocharging.

3. Turbochargers may require removal, prior to engine overhaul, for carbon removal and repair.

4. Engines with reverse rotation have same overhaul times as corresponding normal rotation models.

5. 1200 HOURS: Engines that do not have large main bearing dowels should not be operated more than 1200 hours between overhauls.

1400 HOURS: Engines that have large main bearing dowels may be operated to 1400 hours between overhauls. These include engines with serial numbers L-7100-51A and up; engines which are in compliance with the latest edition of Service Bulletin No. 326; and remanufactured engines shipped after January 26, 1970.

1800 HOURS: Engines that have large main bearing dowels and redesigned camshafts may be operated to 1800 hours between overhauls. These include engines with serial numbers L-9762-51A and up; IO-360-C1E6 engines with serial numbers L-9723-51A and up; LIO-360-C1E6 engines with serial numbers L-524-67A and up; engines that are in compliance with the latest edition of Service Bulletin No. 326 and Service Instruction No. 1263. Remanufactured engines shipped after October 1, 1972 may be operated to 1800 hours between overhauls except those with serial numbers L-2349-51A and L-7852-51A which do not have the redesigned camshaft and must not exceed 1400 hours operating time between overhauls.

6. The reliability and service life of engines can be detrimentally affected if they are repeatedly operated at alternating high and low power applications which cause extreme changes in cylinder temperatures. Flight maneuvers which cause engine overspeed also contribute to abnormal wear characteristics that tend to shorten engine life. These factors must be considered to establish TBO of aerobatic engines; therefore it is the responsibility of the operator to determine the percentage of time the engine is used for aerobatics and establish his own TBO. The maximum recommended is the time specified in this instruction.

7. TIO-540-A series engines with serial numbers L-1880-61 and up, and TIO-540-C series engines with serial numbers L-1754-61 and up, and remanufactured engines built after March 1, 1971 were built with large main bearing dowels and may be operated to 1800 hours. Also, TIO-540-A and -C engines that have been modified to incorporate large main bearing dowels in accordance with latest edition of Service Instruction No. 1225 may be operated to 1800 hours. Engines that do not have this modification incorporated may not exceed 1500 hours between overhauls.

8. VO, TVO and TIVO-540 engines built with P/N 77450 connecting rods as described in the latest edition of Service Bulletin No. 371 may be continued in service to 1200 hours. Engines that do not incorporate this new connecting rod are restricted to 1000 hours for VO-540 models and 900 hours for TVO and TIVO-540. See latest edition of Service Bulletin No. 371 for improved connecting rod assembly.

9. New TIO-541-E engines with serial numbers L-804-59 and up, and remanufactured engines shipped after March 1, 1976 may be continued in service for 1600 hours. Also, remanufactured and overhauled engines which incorporate improved crankcases and cylinder assemblies, as described in the latest edition of Service Bulletin Nos. 334 and 353, may be continued in service for 1600 hours.

10. Some engines in the field have been altered to incorporate an inverted oil system in order to perform aerobatic maneuvers. Whenever this modification is done to an engine, the TBO of the engine must be determined in the same manner listed for AIO and AEIO engines of the same model series.

11. If an engine is being used in "scheduled frequent" type service and accumulates 60 hours or more per month, and has been so operated consistently, add 200 hours to TBO time.

12. To qualify for the 2400 hour TBO, high-compression O-235's must have the increased strength pistons (P/N LW-18729). See latest issue of Service Letter L213.

13. The high-compression O-235-F, -G, and -J series do not have the increased-strength pistons (P/N LW-18729); therefore, they do not qualify for the 2400-hour TBO.

NOTE: Revision "B" revises text on page 1 and chart on page 2. Also, Note 11 of previous revision is deleted; Note 12 becomes Note 11, with new Notes 12 and 13 added.

TELEDYNE CONTINENTAL® AIRCRAFT ENGINE

service bulletin

M86-6 Rev. 1
Supersedes M86-6

Technical Portions Are FAA Approved.

18 April 1986

SUBJECT: **RECOMMENDED OVERHAUL PERIODS FOR ALL TELEDYNE CONTINENTAL MOTORS AIRCRAFT ENGINES.**

MODELS AFFECTED: All

COMPLIANCE: Overhaul

Thousands of hours of operating experience indicate that Teledyne Continental Motors (TCM) aircraft engines, when operated within prescribed limitations, instructions and recommendations, can be operated between overhauls for the number of hours listed in the following table. The overhaul periods listed are recommendations only. They are predicated on the **use of genuine TCM parts,** compliance with all applicable Service Bulletins and ADs, as well as all required preventive maintenance, periodic inspections, manufacturer's specifications, and the determination by a qualified mechanic that the engine is operating normally and is airworthy. The accomplishment of cylinder leakage checks and spectrographic oil analysis may be helpful in making this determination. Any operation beyond these periods is at the operator's discretion and should be based on the inspecting mechanic's evaluation of engine condition and operating environment. Calendar time also affects this condition and should be taken into consideration.

Particular attention should be paid to throttle response, power, smoothness of operation, oil consumption, to the proper use and maintenance of oil and air filters, and adherence to the recommended oil change periods. Emphasis should also be placed on recommended fuel management. These recommended overhaul periods in no way alter TCM's warranty policies.

Strict adherence to the latest Major Overhaul 100% Replacement Parts Service Bulletin is required.

Overhaul periods may vary depending upon the quality of parts and workmanship incorporated into the overhaul of an engine. TCM is unable to provide specific guidelines concerning the effect **non-TCM supplied parts** have on the overhaul periods listed in the following table and makes no recommendation with respect to overhaul periods for engines which have been overhauled, repaired or modified in a manner inconsistent with the specifications, limits and instructions set forth in the latest applicable TCM overhaul manual, parts catalog and service bulletins.

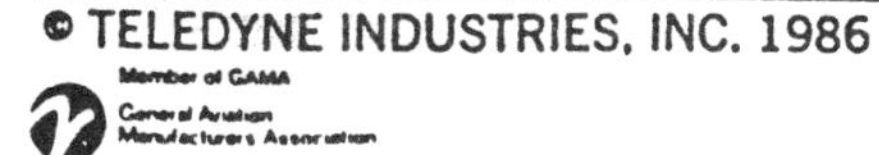

TELEDYNE CONTINENTAL MOTORS
Aircraft Products Division
P. O. Box 90 • Mobile, Alabama 36601

	ENGINE MODEL	OVERHAUL PERIOD
	A65, A75 Series	1800 hours
	C75, C85, C90, O-200 Series	1800 hours
	C125, C145, O-300 Series	1800 hours
	GO-300 Series	1200 hours
	E165, E185, E225 Series	1500 hours
	W670 Series	1000 hours
①	O-470 Series	1500 hours
|	O-470-U	1500 hours
④	O-470-U	2000 hours
①	IO-470 Series	1500 hours
	TSIO-470 Series	1400 hours
	GIO-470A	1000 hours
	IO-346 Series	1500 hours
	IO-360 Series	1500 hours
	IO-360KB	2000 hours
	TSIO/LTSIO-360EB, KB	1800 hours
	TSIO-360FB, GB, LB, MB	1800 hours
	TSIO-360 Series	1400 hours
①	IO-520 Series	1700 hours
①	TSIO-520 Series	1400 hours
	TSIO-520UB, VB, WB, AF, CE	1600 hours
|	TSIO-520NB	1400 hours
③	TSIO-520NB	1600 hours
|	TSIO-520M, P, R	1400 hours
⑤	TSIO-520M, P, R	1600 hours
	TSIO/LTSIO-520AE	2000 hours
	TSIO-520-BE	2000 hours
	GTSIO-520 Series	1200 hours
|	GTSIO-520L	1200 hours
②	GTSIO-520L	1600 hours
	GTSIO-520-M, N	1600 hours
	6-285 Series	1200 hours
	IO-550-B, C	1700 hours

① Excepted are engines employed in aerial dressing, dusting or spraying. For these engines, we recommend a maximum of 1200 hours TBO, or less at operator's discretion.

② Applies to new GTSIO-520L engines having S/N 604701 and subsequent and rebuilt engines having S/N 227809R and subsequent only. Engines may be made eligible for the extended TBO (1600 hours) by installing parts per latest Parts Catalog X30046A and applicable Service Bulletins .

③ Applies to new TSIO-520NB engines having serial numbers 521391 thru 521399, 521400, 521405, 521406, 521411, 521412, 521419 and subsequent and rebuilt TSIO-520-NB engines having serial numbers 234070, 234074 and subsequent only. Engines may be made eligible for the extended TBO (1600 hours) by installing parts per latest Parts Catalog X30580A and applicable Service Bulletins.

④ Applies to new and rebuilt O-470U Spec. 11, 12, 13, 14, 17 and 18.

O-470U engines other than those listed above may be made eligible for the TBO increase (2000 hours) by installing P/N 646267A2 cylinder and valve assembly, P/N 646280 piston, P/N 639565A9 ring set, P/N 646277 lifter in the exhaust position, P/N 643779 oil pump, P/N 643749 oil pump gasket, P/N 643227 oil filter and 2 each P/N 402129 studs (or Cessna supplied oil filter adapter and associated parts). Piston pin P/N 539467 must be replaced (not reused). Crankshaft counterweight pin and plate configuration must conform to current parts catalog X30023A. A log book entry will be required and new Spec No. stamped on engine data plate (Refer to Engine Spec. List below).

ORIGINAL SPEC. #	NEW SPEC. #
O470U1	13
O470U2	14
O470U3	17
O470U4	18
O470U5	17
O470U6	18

⑤ Applies to new and rebuilt TSIO-520-M Spec. 4, 6, 7 and 8; TSIO-520-P Spec. 5 and 6; and TSIO-520-R Spec. 7, 9, 10 and 11.

TSIO-520-M, P & R engines other than those listed above may be eligible for the TBO increase (1600 hours) by installing cylinder & valve assemblies P/N 646657A1, piston P/N 648033, ring set P/N 642602A2, valve lifters P/N 646277 in both intake and exhaust positions, throttle body P/N 649185A1, pressurized magneto kit EQ6583, oil pump assembly P/N 643717-1, P/N 643749 oil pump gasket, and oil filter P/N 643227 (or Cessna supplied oil filter adapter and associated parts).

To install new oil pump remove one (1) each P/N 402159 and P/N 402157 stud. Replace stud P/N 401852 with stud P/N 402129 and install spacer P/N 646582-1.35 and P/N 646582-2.00 on existing studs after oil pump is installed. A log book entry will be required and new Spec. No. stamped on engine data plate (Refer to Engine Spec. List below).

ORIGINAL SPEC. #	NEW SPEC. #
TSIO520P1	5
TSIO520P2	6
TSIO520P3	6
TSIO520M1	6
TSIO520M2	7
TSIO520M3	7
TSIO520R1	9
TSIO520R3	10
TSIO520R4	9
TSIO520R5	10
TSIO520R6	11

Service Newsletter

March 28, 1986 SNL86-11

TITLE

CESSNA AIRCRAFT SERIAL NUMBER LISTING

TO

CESSNA DISTRIBUTORS, CATEGORY I THRU CATEGORY IV DEALERS, CARAVAN I AND II REPRESENTATIVES/SERVICE STATIONS, AG DEALERS AND CPC'S

DISCUSSION

Attached for your use is a copy of the latest aircraft serial number listing for all Cessna propeller aircraft.

The listing shows aircraft serial numbers, gross weight/maximum takeoff weight, engine model and horsepower for each model year from 1946 to the current 1986 models.

The serial listing is a convenient "quick reference" document, but should be used as general information only.

* * * * * * * *

Reprinted by permission of
Dean Humphrey,
Director of Public Relations
Cessna Aircraft
April 8, 1988

FAA MODEL-SERIAL CODE	MODEL	YEAR	SERIALS		GROSS/T.O. WT.	ENGINE	
			Beginning	Ending		Model	Horsepower
120/14-02	120-140	1946	8000	11846	1450	C-85	85
140/16-02	120-140	1947	11847	14370	1450	C-85	85
	120-140	1948 & 1949	14371	15075	1450	C-85	85
16-04	140A	1949 - 1951	15200	15724	1500	C-85 & C-90	85 & 90
18-02	150	1959	17001	17683	1500	0-200-A	100
18-02	150	1960	17684	17999	1500	0-200-A	100
18-02			59001	59018	1500	0-200-A	100
18-04	150A	1961	15059019	15059350	1500	0-200-A	100
18-06	150B	1962	15059351	15059700	1500	0-200-A	100
18-08	150C	1963	15059701	15060087	1500	0-200-A	100
18-10	150D	1964	15060088	15060772	1600	0-200-A	100
18-12	150E	1965	15060773	15061532	1600	0-200-A	100
18-14	150F	1966	15061533	15064532	1600	0-200-A	100
18-16	150G	1967	15064533	15067198	1600	0-200-A	100
18-18	150H	1968	15067199	15069308	1600	0-200-A	100
18-20	150J	1969	15069309	15071128	1600	0-200-A	100
18-22	150K	1970	15071129	15072003	1600	0-200-A	100
18-26	150L	1971	15072004	15072628	1600	0-200-A	100
18-26	150L	1972	15072629	15073658	1600	0-200-A	100
18-26	150L	1973	15073659	15074850	1600	0-200-A	100
18-26	150L	1974	15074851	15075781	1600	0-200-A	100
18-30	150M	1975	15075782	15077005	1600	0-200-A	100
18-30	150M	1976	15077006	15078505	1600	0-200-A	100
18-30	150M	1977	15078506	15079405	1600	0-200-A	100
18-24	150 AEROBAT (A150K)	1970	A1500001	A1500226	1600	0-200-A	100
18-28	150 AEROBAT (A150L)	1971	A1500227	A1500276	1600	0-200-A	100
18-28	150 AEROBAT (A150L)	1972	A1500277	A1500342	1600	0-200-A	100
18-28	150 AEROBAT (A150L)	1973	A1500343	A1500429	1600	0-200-A	100
18-28	150 AEROBAT (A150L)	1974	A1500430	A1500523	1600	0-200-A	100
18-31	150 AEROBAT (A150M)	1975	A1500524	A1500609	1600	0-200-A	100
18-31	150 AEROBAT (A150M)	1976	A1500610	A1500684	1600	0-200-A	100
18-31	150 AEROBAT (A150M)	1977	A1500685	A1500734	1600	0-200-A	100
18-35	152	1978	15279406	15282031	1670	0-235-L2C	110
18-35	152	1979	15282032	15283591	1670	0-235-L2C	110
18-35	152	1980	15283592	15284541	1670	0-235-L2C	110
18-35	152	1981	15284542	15285161	1670	0-235-L2C	110
18-35	152	1982	15285162	15285594	1670	0-235-L2C	110
18-35	152	1983	15285595	15285833	1670	0-235-N2C	108
18-35	152	1984	15285834	15285939	1670	0-235-N2C	108
18-35	152	1985	15285940	15286033	1670	0-235-N2C	108
18-36	A152 AEROBAT	1978	A1520735	A1520808	1670	0-235-L2C	110
18-36	A152 AEROBAT	1979	A1520809	A1520878	1670	0-235-L2C	110
18-36	A152 AEROBAT	1980	A1520879	A1520943	1670	0-235-L2C	110
18-36	A152 AEROBAT	1981	A1520944	A1520983	1670	0-235-L2C	110
18-36	A152 AEROBAT	1982	A1520984	A1521014	1670	0-235-L2C	110
18-36	A152 AEROBAT	1983	A1521015	A1521025	1670	0-235-N2C	108
18-36	A152 AEROBAT	1984	A1521026	A1521027	1670	0-235-N2C	108
18-36	A152 AEROBAT	1985	A1521028	A1521049	1670	0-235-N2C	108
23-02	170	1948	18000	18729	2200	C-145	145
23-04	170A	1949	18730	19199	2200	C-145	145
23-04	170A	1950 & 1951	19200	20266	2200	C-145	145
23-06	170B	1952	20267	20999	2200	C-145	145
23-06			25000	25372	2200	C-145	145
23-06	170B	1953	25373	26038	2200	C-145	145
23-06	170B	1954	26039	26504	2200	C-145	145
23-06	170B	1955	26505	26995	2200	0-300-A	145
23-06	170B	1956	26996	27169	2200	0-300-A	145
24-02	172	1956	28000	29174	2200	0-300-A	145
24-02	172	1957	29175	29999	2200	0-300-A	145
24-02			36000	36215	2200	0-300-A	145
24-02	172	1958	36216	36965	2200	0-300-A	145
24-02	172	1959	36966	36999	2200	0-300-A	145
24-02			46001	46754	2200	0-300-A	145
24-04	172A	1960	46755	47746	2200	0-300-C	145
24-06	172B/SKYHAWK	1961	17247747	17248734	2200	0-300-C & D	145
24-08	172C/SKYHAWK	1962	17248735	17249544	2250	0-300-C & D	145
24-10	172D/SKYHAWK	1963	17249545	17250572	2300	0-300-C & D	145

FAA MODEL-SERIAL CODE	MODEL	YEAR	SERIALS Beginning	Ending	GROSS/T.O. WT.	ENGINE Model	Horsepower
24-12	172E/SKYHAWK	1964	17250573	17251822	2300	0-300-C & D	145
24-14	172F/SKYHAWK	1965	17251823	17253392	2300	0-300-C & D	145
24-20	172G/SKYHAWK	1966	17253393	17254892	2300	0-300-C & D	145
24-24	172H/SKYHAWK	1967	17254893	17256512	2300	0-300-C & D	145
24-26	172I/SKYHAWK	1968	17256513	17257161	2300	0-320-E2D	150
24-30	172K/SKYHAWK	1969	17257162	17258486	2300	0-320-E2D	150
24-30	172K/SKYHAWK	1970	17258487	17259223	2300	0-320-E2D	150
24-32	172L/SKYHAWK	1971	17259224	17259903	2300	0-320-E2D	150
24-32	172L/SKYHAWK	1972	17259904	17260758	2300	0-320-E2D	150
24-18	172M/SKYHAWK	1973	17260759	17261898	2300	0-320-E2D	150
24-18	172M/SKYHAWK	1974	17261899	17263458	2300	0-320-E2D	150
24-18	172M/SKYHAWK	1975	17263459	17265684	2300	0-320-E2D	150
24-18	SKYHAWK (172M)	1976	17265685	17267584	2300	0-320-E2D	150
24-34	SKYHAWK 100 (172N)	1977	17267585	17269309	2300	0-320-H2AD	160
24-34	SKYHAWK (172N)	1978	17269310	17271034	2300	0-320-H2AD	160
24-34	SKYHAWK (172N)	1979	17271035	17272884	2300	0-320-H2AD	160
24-34	SKYHAWK (172N)	1980	17272885	17274009	2300	0-320-H2AD	160
24-36	SKYHAWK (172P)	1981	17274010	17275034	2400	0-320-D2J	160
24-36	SKYHAWK (172P)	1982	17275035	17275759	2400	0-320-D2J	160
24-36	SKYHAWK (172P)	1983	17275760	17276079	2400	0-320-D2J	160
24-36	SKYHAWK (172P)	1984	17276080	17276259	2400	0-320-D2J	160
24-36	SKYHAWK (172P)	1985	17276260	17276516	2400	0-320-D2J	160
24-36	SKYHAWK (172P)	1986	17276517		2400	0-320-D2J	160
24-37	CUTLASS (172Q)	1983	17275869	17276079	2550	0-360-A4N	180
24-37	CUTLASS (172Q)	1984	17276080	17276259	2550	0-360-A4N	180
24-38	CUTLASS RG (172 RG)	1980	172RG0001	172RG0570	2650	0-360-F1A6	180
24-38	CUTLASS RG (172 RG)	1981	172RG0571	172RG0890	2650	0-360-F1A6	180
24-38	CUTLASS RG (172 RG)	1982	172RG0891	172RG1099	2650	0-360-F1A6	180
24-38	CUTLASS RG (172 RG)	1983	172RG1100	172RG1144	2650	0-360-F1A6	180
24-38	CUTLASS RG (172RG)	1984	172RG1145	172RG1177	2650	0-360-F1A6	180
24-38	CUTLASS RG (172RG)	1985	172RG1178	172RG1191	2650	0-360-F1A6	180
24-31	HAWK XP (R172K)	1977	R1722000	R1722724	2550	IO-360-K	195
24-31	HAWK XP (R172K)	1978	R1722725	R1722929	2550	IO-360-K	195
24-31	HAWK XP (R172K)	1979	R1722930	R1723199	2550	IO-360-K & KB	195
24-31	HAWK XP (R172K)	1980	R1723200	R1723399	2550	IO-360-KB	195
24-31	HAWK XP (R172K)	1981	R1723400	R1723454	2550	IO-360-KB	195
25-02	175	1958	55001	55703	2350	GO-300-A	175
25-02	175	1959	55704	56238	2350	GO-300-A	175
25-04	175A/SKYLARK	1960	56239	56777	2350	GO-300-C	175
25-06	175B/SKYLARK	1961	17556778	17557002	2350	GO-300-C & D	175
25-08	SKYLARK (175C)	1962	17557003	17557119	2450	GO-300-E	175
22-02	P172/SKYHAWK POWERMATIC	1963	P17257120	P17257188	2500	GO-300-E	175
37-04	177/CARDINAL	1968	17700001	17701164	2350	0-320-E2D	150
37-06	177A/CARDINAL	1969	17701165	17701370	2500	0-360-A2F	180
37-08	177B/CARDINAL	1970	17701371	17701530	2500	0-360-A1F6	180
37-08	177B/CARDINAL	1971	17701531	17701633	2500	0-360-A1F6	180
37-08	177B/CARDINAL	1972	17701634	17701773	2500	0-360-A1F6	180
37-08	177B/CARDINAL	1973	17701774	17701973	2500	0-360-A1F6D	180
37-08	177B/CARDINAL	1974	17701974	17702123	2500	0-360-A1F6D	180
37-08	177B/CARDINAL	1975	17702124	17702313	2500	0-360-A1F6D	180
37-08	CARDINAL (177B)	1976	17702314	17702522	2500	0-360-A1F6D	180
37-08	CARDINAL (177B)	1977	17702523	17702672	2500	0-360-A1F6D	180
37-08	CARDINAL CLASSIC (177B)	1978	17702673	17702752	2500	0-360-A1F6D	180
37-09	CARDINAL RG (177 RG)	1971	177RG0001	177RG0212	2800	IO-360-A1B6	200
37-09	CARDINAL RG (177 RG)	1972	177RG0213	177RG0282	2800	IO-360-A1B6	200
37-09	CARDINAL RG (177 RG)	1973	177RG0283	177RG0432	2800	IO-360-A1B6D	200
37-09	CARDINAL RG (177 RG)	1974	177RG0433	177RG0592	2800	IO-360-A1B6D	200
37-09	CARDINAL RG (177 RG)	1975	177RG0593	177RG0787	2800	IO-360-A1B6D	200
37-09	CARDINAL RG (177 RG)	1976	177RG0788	177RG1051	2800	IO-360-A1B6D	200
37-09	CARDINAL RG (177 RG)	1977	177RG1052	177RG1266	2800	IO-360-A1B6D	200
37-09	CARDINAL RG (177 RG)	1978	177RG1267	177RG1366	2800	IO-360-A1B6D	200
26-02	180	1953	30000	30639	2550	0-470-A	225
26-02	180	1954	30640	31259	2550	0-470-A & J	225

FAA MODEL-SERIAL CODE	MODEL	YEAR	SERIALS Beginning	Ending	GROSS/T.O. WT.	ENGINE Model	Horsepower
26-02	180	1955	31260	32150	2550	0-470-J	225
26-02	180	1956	32151	32661	2550	0-470-K	230
26-04	180A	1957	32662	32999	2650	0-470-K	230
26-04			50000	50105	2650	0-470-K	230
26-04	180A	1958	50106	50355	2650	0-470-K	230
26-06	180B	1959	50356	50661	2650	0-470-K	230
26-08	180C	1960	50662	50911	2650	0-470-L	230
26-10	180D	1961	18050912	18051063	2650	0-470-L	230
26-12	180E	1962	18051064	18051183	2650	0-470-R	230
26-14	180F	1963	18051184	18051312	2650	0-470-R	230
26-16	180G	1964	18051313	18051445	2800	0-470-R	230
26-18	180H	1965	18051446	18051607	2800	0-470-R	230
26-18	180H	1966	18051608	18051774	2800	0-470-R	230
26-18	180H	1967	18051775	18051875	2800	0-470-R	230
26-18	180H	1968	18051876	18051993	2800	0-470-R	230
26-18	SKYWAGON 180 (H)	1969	18051994	18052103	2800	0-470-R	230
26-18	SKYWAGON 180 (H)	1970	18052104	18052175	2800	0-470-R	230
26-18	SKYWAGON 180 (H)	1971	18052176	18052221	2800	0-470-R	230
26-18	SKYWAGON 180 (H)	1972	18052222	18052284	2800	0-470-R	230
26-22	SKYWAGON 180 (J)	1973	18052285	18052384	2800	0-470-R	230
26-22	SKYWAGON 180 (J)	1974	18052385	18052500	2800	0-470-R	230
26-22	SKYWAGON 180 (J)	1975	18052501	18052620	2800	0-470-S	230
26-22	180 SKYWAGON (J)	1976	18052621	18052770	2800	0-470-S	230
26-24	180 SKYWAGON (K)	1977	18052771	18052905	2800	0-470-U	230
26-24	180 SKYWAGON (K)	1978	18052906	18053000	2800	0-470-U	230
26-24	180 SKYWAGON (K)	1979	18053001	18053115	2800	0-470-U	230
26-24	180 SKYWAGON (K)	1980	18053116	18053167	2800	0-470-U	230
26-24	180 SKYWAGON (K)	1981	18053168	18053203	2800	0-470-U	230
27-02	182	1956	33000	33842	2550	0-470-L	230
27-04	182A	1957	33843	34753	2650	0-470-L	230
27-04	182A/SKYLANE	1958	34754	34999	2650	0-470-L	230
27-04			51001	51556	2650	0-470-L	230
27-06	182B/SKYLANE	1959	51557	52358	2650	0-470-L	230
27-08	182C/SKYLANE	1960	52359	53007	2650	0-470-L	230
27-10	182D/SKYLANE	1961	18253008	18253598	2650	0-470-L	230
27-12	182E/SKYLANE	1962	18253599	18254423	2800	0-470-R	230
27-14	182F/SKYLANE	1963	18254424	18255058	2800	0-470-R	230
27-16	182G/SKYLANE	1964	18255059	18255844	2800	0-470-R	230
27-18	182H/SKYLANE	1965	18255845	18256684	2800	0-470-R	230
27-22	182J/SKYLANE	1966	18256685	18257625	2800	0-470-R	230
27-24	182K/SKYLANE	1967	18257626	18258505	2800	0-470-R	230
27-26	182L/SKYLANE	1968	18258506	18259305	2800	0-470-R	230
27-28	182M/SKYLANE	1969	18259306	18260055	2800	0-470-R	230
27-30	182N/SKYLANE	1970	18260056	18260445	2950	0-470-R	230
27-30	182N/SKYLANE	1971	18260446	18260825	2950	0-470-R	230
58-16	182P/SKYLANE	1972	18260826	18261425	2950	0-470-R	230
58-16	182P/SKYLANE	1973	18261426	18262465	2950	0-470-R	230
58-16	182P/SKYLANE	1974	18262466	18263475	2950	0-470-R	230
58-16	182P/SKYLANE	1975	18263476	18264295	2950	0-470-S	230
58-16	SKYLANE (182P)	1976	18264296	18265175	2950	0-470-S	230
27-32	SKYLANE (182Q)	1977	18265176	18265965	2950	0-470-U	230
27-32	SKYLANE (182Q)	1978	18265966	18266590	2950	0-470-U	230
27-32	SKYLANE (182Q)	1979	18266591	18267300	2950	0-470-U	230
27-32	SKYLANE (182Q)	1980	18267301	18267715	2950	0-470-U	230
27-31	SKYLANE (182R)	1981	18267716	18268055	3100	0-470-U	230
27-36	TURBO SKYLANE (T182)		18267716	18268055	3100	0-540-L3C5D	235
27-31	SKYLANE (182R)	1982	18268056	18268293	3100	0-470-U	230
27-36	TURBO SKYLANE (T182)		18268056	18268293	3100	0-540-L3C5D	235
27-31	SKYLANE (182R)	1983	18268294	18268368	3100	0-470-U	230
27-36	TURBO SKYLANE (T182)		18268294	18268368	3100	0-540-L3C5D	235
27-31	SKYLANE (182R)	1984	18268369	18268434	3100	0-470-U	230
27-36	TURBO SKYLANE (T182)		18268369	18268434	3100	0-540-L3C5D	235
27-31	SKYLANE (182R)	1985	18268435	18268541	3100	0-470-U	230
27-36	TURBO SKYLANE (T182)		18268435	18268541	3100	0-540-L3C5D	235
27-31	SKYLANE (182R)	1986	18268542		3100	0-470-U	230
27-36	TURBO SKYLANE (T182)		18268542		3100	0-540-L3C5D	235

FAA MODEL-SERIAL CODE	MODEL	YEAR	SERIALS Beginning	SERIALS Ending	GROSS/T.O. WT.	ENGINE Model	ENGINE Horsepower
27-34	SKYLANE RG (R182)	1978	R18200001	R18200583	3100	0-540-J3C5D	235
27-34	SKYLANE RG (R182)	1979	R18200584	R18201313	3100	0-540-J3C5D	235
27-35	TURBO SKYLANE RG (TR182)		R18200584	R18201313	3100	0-540-L3C5D	235
27-34	SKYLANE RG (R182)	1980	R18201314	R18201628	3100	0-540-J3C5D	235
27-35	TURBO SKYLANE RG (TR182)		R18201314	R18201628	3100	0-540-L3C5D	235
27-34	SKYLANE RG (R182)	1981	R18201629	R18201798	3100	0-540-J3C5D	235
27-35	TURBO SKYLANE RG (TR182)		R18201629	R18201798	3100	0-540-L3C5D	235
27-34	SKYLANE RG (R182)	1982	R18201799	R18201928	3100	0-540-J3C5D	235
27-35	TURBO SKYLANE RG (TR182)		R18201799	R18201928	3100	0-540-L3C5D	235
27-34	SKYLANE RG (R182)	1983	R18201929	R18201973	3100	0-540-J3C5D	235
27-35	TURBO SKYLANE RG (TR182)		R18201929	R18201973	3100	0-540-L3C5D	235
27-34	SKYLANE RG (R182)	1984	R18201974	R18201999	3100	0-540-J3C5D	235
27-35	TURBO SKYLANE RG (TR182)		R18201974	R18201999	3100	0-540-L3C5D	235
27-34	SKYLANE RG (R182)	1985	R18202000	R18202031	3100	0-540-J3C5D	235
27-35	TURBO SKYLANE RG (TR182)		R18202000	R18202031	3100	0-540-L3C5D	235
27-34	SKYLANE RG (R182)	1986	R18202032		3100	0-540-J3C5D	235
27-35	TURBO SKYLANE RG (TR182)		R18202032		3100	0-540-L3C5D	235

FAA MODEL-SERIAL CODE	MODEL	YEAR	SERIALS Beginning	SERIALS Ending	GROSS/T.O. WT.	ENGINE Model	ENGINE Horsepower
28-02	185 SKYWAGON	1961	185-0001	185-0237	3200	IO-470-F	260
28-04	185 SKYWAGON (A)	1962	185-0238	185-0512	3200	IO-470-F	260
28-06	185 SKYWAGON (B)	1963	185-0513	185-0653	3200	IO-470-F	260
28-08	185 SKYWAGON (C)	1964	185-0654	185-0776	3200	IO-470-F	260
28-12	185 SKYWAGON (D)	1965	185-0777	185-0967	3200	IO-470-F	260
28-16	185 SKYWAGON (E)	1966	185-0968	185-1149	3300	IO-470-F	260
28-18	185 SKYWAGON (E)	1966	185-0968	185-1149	3350	IO-520-D	300*
28-18	185 SKYWAGON (A185E)	1967	185-1150	185-1300	3350	IO-520-D	300*
28-18	185 SKYWAGON (A185E)	1968	185-1301	185-1447	3350	IO-520-D	300*
28-18	SKYWAGON 185 (A185E)	1969	185-1448	185-1599	3350	IO-520-D	300*
28-18	SKYWAGON 185 (A185E)	1970	18501600	18501832	3350	IO-520-D	300*
28-18	SKYWAGON 185 (A185E)	1971	18501833	18501934	3350	IO-520-D	300*
28-18	SKYWAGON 185 (A185E) & AGcarryall	1972	18501935	18502090	3350	IO-520-D	300*
28-21	SKYWAGON 185 (A185F) & AGcarryall	1973	18502091	18502310	3350	IO-520-D	300*
28-21	SKYWAGON 185 (A185F) & AGcarryall	1974	18502311	18502565	3350	IO-520-D	300*
28-21	SKYWAGON 185 (A185F) & AGcarryall	1975	18502566	18502838	3350	IO-520-D	300*
28-21	185 SKYWAGON (A185F) & AGcarryall	1976	18502839	18503153	3350	IO-520-D	300*
28-21	185 SKYWAGON (A185F) & AGcarryall	1977	18503154	18503458	3350	IO-520-D	300*
28-21	185 SKYWAGON (A185F) & AGcarryall	1978	18503459	18503683	3350	IO-520-D	300*
28-21	185 SKYWAGON (A185F) & AGcarryall	1979	18503684	18503938	3350	IO-520-D	300*
28-21	185 SKYWAGON (A185F)	1980	18503939	18504138	3350	IO-520-D	300*
28-21	185 SKYWAGON (A185F)	1981	18504139	18504328	3350	IO-520-D	300*
28-21	185 SKYWAGON (A185F)	1982	18504329	18504394	3350	IO-520-D	300*
28-21	185 SKYWAGON (A185F)	1983	18504395	18504415	3350	IO-520-D	300*
28-21	185 SKYWAGON (A185F)	1984	18504416	18504424	3350	IO-520-D	300*
28-21	185 SKYWAGON (A185F)	1985	18504425	18504448	3350	IO-520-D	300*

FAA MODEL-SERIAL CODE	MODEL	YEAR	SERIALS Beginning	SERIALS Ending	GROSS/T.O. WT.	ENGINE Model	ENGINE Horsepower
30-02	AG WAGON 230 (188)	1966 & 1967	188-0001	188-0317	3800**	0-470-R	230
30-04	AG WAGON 300 (A188)		188-0001	188-0317	4000**	IO-520-D	300*
30-06	AG WAGON "A" (188)	1968 & 1969	188-0318	188-0572	3800**	0-470-R	230
30-08	300 HP ENGINE OPTION (A188)		188-0318	188-0572	4000**	IO-520-D	300*
30-06	AG WAGON "B" (188A)	1970 & 1971	18800573	18800832	3800**	0-470-R	230
30-08	300 HP ENGINE OPTION (A188A)		18800573	18800832	4000**	IO-520-D	300*
30-07	AG PICKUP (188B)	1972	18800833	18801040	3800**	0-470-R	230
30-05	AG WAGON "C" (A188B)		18800833	18801040	4000**	IO-520-D	300*
30-05	AG TRUCK (A188B)		18800967T	18801040T	4000**	IO-520-D	300*
30-07	AG PICKUP (188B)	1973	18801041	18801374	3800**	0-470-R	230
30-05	AG WAGON (A188B)		18801041	18801374	4000**	IO-520-D	300*
30-05	AG TRUCK (A188B)		18801041T	18801374T	4000**	IO-520-D	300*
30-07	AG PICKUP (188B)	1974	18801375	18801824	3800**	0-470-R	230
30-05	AG WAGON (A188B)		18801375	18801824	4000**	IO-520-D	300*
30-05	AG TRUCK (A188B)		18801375T	18801824T	4200**	IO-520-D	300*
30-07	AG PICKUP (188B)	1975	18801825	18802348	3800**	0-470-S	230
30-05	AG WAGON (A188B)		18801825	18802348	4000**	IO-520-D	300*
30-05	AG TRUCK (A188B)		18801825T	18802348T	4200**	IO-520-D	300*
30-05	AG WAGON (A188B)	1976	18802349	18802745	4000**	IO-520-D	300*
30-05	AG TRUCK (A188B)		18802349T	18802745T	4200**	IO-520-D	300*

**RESTRICTED CATEGORY WEIGHT
NORMAL CATEGORY IS 3300 LBS.

*5 MIN. TAKEOFF RATING CONTINUOUS
MAXIMUM RATING IS 285 HORSEPOWER.

FAA MODEL-SERIAL CODE	MODEL	YEAR	SERIALS Beginning	SERIALS Ending	GROSS/T.O. WT.	ENGINE Model	ENGINE Horsepower
30-05	AG WAGON (A188B)	1977	18802746	18803046	4000**	IO-520-D	300*
30-05	AG TRUCK (A188B)		18802746T	18803046T	4200**	IO-520-D	300*
30-05	AG WAGON (A188B)	1978	18803047	18803296	4000**	IO-520-D	300*
30-05	AG TRUCK (A188B)		18803047T	18803296T	4200**	IO-520-D	300*
30-05	AG WAGON (A188B)	1979	18803297	18803521	4000**	IO-520-D	300*
30-05	AG TRUCK (A188B)		18803297T	18803521T	4200**	IO-520-D	300*
30-12	AG HUSKY (T188C)		T18803307T	T18803521T	4400**	TSIO-520-T	310*
30-05	AG WAGON (A188B)	1980	18803522	18803721	4000**	IO-520-D	300*
30-05	AG TRUCK (A188B)		18803522T	18803721T	4200**	IO-520-D	300*
30-12	AG HUSKY (T188C)		T18803522T	T18803721T	4400	TSIO-520-T	310*
30-05	AG WAGON (A188B)	1981	18803722	18803856	4000**	IO-520-D	300*
30-05	AG TRUCK (A188B)		18803722T	18803856T	4200**	IO-520-D	300*
30-12	AG HUSKY (T188C)		T18803722T	T18803856T	4400	TSIO-520-T	310*
30-05	AG TRUCK (A188B)	1982	18803857T	18803926T	4200**	IO-520-D	300*
30-12	AG HUSKY (T188C)		T18803857T	T18803926T	4400	TSIO-520-T	310*
30-05	AG TRUCK (A188B)	1983	18803927T	18803968T	4200**	IO-520-D	300*
30-12	AG HUSKY (T188C)		T18803927T	T18803968T	4400	TSIO-520-T	310*
190/29-02	190-195 SERIES	1948-1953	7001	7999	3350	R-755-9 (195-195A)	245***
195/31-02			16000	16183		R-755A-2 (195)	300
195A/31-10						W-670-23 (190)	240
195B/31-12						R-755B-2 (195B)	275
32-02	205	1963	205-0001	205-0480	3300	IO-470-S	260
32-04	205A	1964	205-0481	205-0577	3300	IO-470-S	260
33-02	SUPER SKYWAGON	1964	206-0001	206-0275	3300	IO-520-A	285
33-06	SUPER SKYWAGON	1965	U206-0276	U206-0437	3300	IO-520-A	285
33-16	SUPER SKYWAGON (A)	1966	U206-0438	U206-0656	3600	IO-520-A	285
33-18	TURBO-SYSTEM Engine Option		U206-0438	U206-0656	3600	TSIO-520-C	285
33-22	SUPER SKYWAGON (B)	1967	U206-0657	U206-0914	3600	IO-520-F	300*
33-34	TURBO-SYSTEM Engine Option		U206-0657	U206-0914	3600	TSIO-520-C	285
33-32	SUPER SKYWAGON (C)	1968	U206-0915	U206-1234	3600	IO-520-F	300*
33-24	TURBO-SYSTEM Engine Option		U206-0915	U206-1234	3600	TSIO-520-C	285
33-42	SKYWAGON 206 (D)	1969	U206-1235	U206-1444	3600	IO-520-F	300*
33-44	TURBO-SKYWAGON 206		U206-1235	U206-1444	3600	TSIO-520-C	285
33-50	SKYWAGON 206 (E)	1970	U20601445	U20601587	3600	IO-520-F	300*
33-52	TURBO-SKYWAGON 206		U20601445	U20601587	3600	TSIO-520-C	285
33-50	STATIONAIR (E)	1971	U20601588	U20601700	3600	IO-520-F	300*
33-52	TURBO STATIONAIR		U20601588	U20601700	3600	TSIO-520-C	285
33-33	STATIONAIR (F)	1972	U20601701	U20601874	3600	IO-520-F	300*
33-53	TURBO STATIONAIR		U20601701	U20601874	3600	TSIO-520-C	285
33-33	STATIONAIR (F)	1973	U20601875	U20602199	3600	IO-520-F	300*
33-53	TURBO STATIONAIR		U20601875	U20602199	3600	TSIO-520-C	285
33-33	STATIONAIR (F)	1974	U20602200	U20602579	3600	IO-520-F	300*
33-53	TURBO STATIONAIR		U20602200	U20602579	3600	TSIO-520-C	285
33-33	STATIONAIR (F)	1975	U20602580	U20603020	3600	IO-520-F	300*
33-53	TURBO STATIONAIR		U20602580	U20603020	3600	TSIO-520-C	285
33-33	STATIONAIR (F)	1976	U20603021	U20603521	3600	IO-520-F	300*
33-53	TURBO STATIONAIR		U20603021	U20603521	3600	TSIO-520-C	285
33-56	STATIONAIR (G)	1977	U20603522	U20604074	3600	IO-520-F	300*
33-57	TURBO STATIONAIR		U20603522	U20604074	3600	TSIO-520-M	310*
33-56	STATIONAIR 6 (G)	1978	U20604075	U20604649	3600	IO-520-F	300*
33-57	TURBO STATIONAIR 6		U20604075	U20604649	3600	TSIO-520-M	310*
33-56	STATIONAIR 6 (G)	1979	U20604650	U20605309	3600	IO-520-F	300*
33-57	TURBO STATIONAIR 6		U20604650	U20605309	3600	TSIO-520-M	310*
33-56	STATIONAIR 6 (G)	1980	U20605310	U20605919	3600	IO-520-F	300*
33-57	TURBO STATIONAIR 6		U20605310	U20605919	3600	TSIO-520-M	310*
33-56	STATIONAIR 6 (G)	1981	U20605920	U20606439	3600	IO-520-F	300*
33-57	TURBO STATIONAIR 6		U20605920	U20606439	3600	TSIO-520-M	310*
33-56	STATIONAIR 6 (G)	1982	U20606440	U20606699	3600	IO-520-F	300*
33-57	TURBO STATIONAIR 6		U20606440	U20606699	3600	TSIO-520-M	310*
33-56	STATIONAIR 6 (G)	1983	U20606700	U20606788	3600	IO-520-F	300*
33-57	TURBO STATIONAIR 6		U20606700	U20606788	3600	TSIO-520-M	310*
33-56	STATIONAIR 6 (G)	1984	U20606789	U20606846	3600	IO-520-F	300*
33-57	TURBO STATIONAIR 6		U20606789	U20606846	3600	TSIO-520-M	310*
33-56	STATIONAIR 6 (G)	1985	U20606847	U20606920	3600	IO-520-F	300*
33-57	TURBO STATIONAIR 6		U20606847	U20606920	3600	TSIO-520-M	310*
33-56	STATIONAIR 6 (G)	1986	U20606921		3600	IO-520-F	300*
33-57	TURBO STATIONAIR 6		U20606921		3600	TSIO-520-M	310*

**RESTRICTED CATEGORY WEIGHT. NORMAL CATEGORY IS 3300 LBS.

***TAKEOFF RATING. CONTINUOUS MAXIMUM RATING IS 225 HORSEPOWER.

*5 MIN. TAKEOFF RATING. CONTINUOUS MAXIMUM RATING IS 285 HORSEPOWER

FAA MODEL-SERIAL CODE	MODEL	YEAR	SERIALS Beginning	SERIALS Ending	GROSS/T.O. WT.	ENGINE Model	ENGINE Horsepower
33-04	SUPER SKYLANE	1965	P206-0001	P206-0160	3300	IO-520-A	285
33-08	SUPER SKYLANE (A)	1966	P206-0161	P206-0306	3600	IO-520-A	285
33-10	TURBO-SYSTEM Engine Option		P206-0161	P206-0306	3600	TSIO-520-C	285
33-09	SUPER SKYLANE (B)	1967	P206-0307	P206-0419	3600	IO-520-A	285
33-11	TURBO-SYSTEM Engine Option		P206-0307	P206-0419	3600	TSIO-520-C	285
33-12	SUPER SKYLANE (C)	1968	P206-0420	P206-0519	3600	IO-520-A	285
33-13	TURBO-SYSTEM Engine Option		P206-0420	P206-0519	3600	TSIO-520-C	285
33-38	SUPER SKYLANE (D)	1969	P206-0520	P206-0603	3600	IO-520-A	285
33-40	TURBO-SYSTEM Engine Option		P206-0520	P206-0603	3600	TSIO-520-C	285
33-46	SUPER SKYLANE (E)	1970	P20600604	P20600647	3600	IO-520-A	285
33-48	TURBO SUPER SKYLANE		P20600604	P20600647	3600	TSIO-520-C	285
36-02	SKYWAGON 207	1969	20700001	20700148	3800	IO-520-F	300*
36-12	TURBO SKYWAGON 207 Engine Option		20700001	20700148	3800	TSIO-520-G	300*
36-02	SKYWAGON 207	1970	20700149	20700190	3800	IO-520-F	300*
36-12	TURBO SKYWAGON 207 Engine Option		20700149	20700190	3800	TSIO-520-G	300*
36-02	SKYWAGON 207	1971	20700191	20700205	3800	IO-520-F	300*
36-12	TURBO SKYWAGON 207 Engine Option		20700191	20700205	3800	TSIO-520-G	300*
36-02	SKYWAGON 207	1972	20700206	20700215	3800	IO-520-F	300*
36-12	TURBO SKYWAGON 207 Engine Option		20700206	20700215	3800	TSIO-520-G	300*
36-02	SKYWAGON 207	1973	20700216	20700227	3800	IO-520-F	300*
36-12	TURBO SKYWAGON 207 Engine Option		20700216	20700227	3800	TSIO-520-G	300*
36-02	SKYWAGON 207	1974	20700228	20700267	3800	IO-520-F	300*
36-12	TURBO SKYWAGON 207		20700228	20700267	3800	TSIO-520-G	300*
36-02	SKYWAGON 207	1975	20700268	20700314	3800	IO-520-F	300*
36-12	TURBO SKYWAGON 207		20700268	20700314	3800	TSIO-520-G	300*
36-02	207 SKYWAGON	1976	20700315	20700362	3800	IO-520-F	300*
36-12	207 TURBO SKYWAGON		20700315	20700362	3800	TSIO-520-G	300*
36-04	207 SKYWAGON (A)	1977	20700363	20700414	3800	IO-520-F	300*
36-14	207 TURBO SKYWAGON		20700363	20700414	3800	TSIO-520-M	310*
36-04	STATIONAIR 7 (A)	1978	20700415	20700482	3800	IO-520-F	300*
36-14	TURBO STATIONAIR 7		20700415	20700482	3800	TSIO-520-M	310*
36-04	STATIONAIR 7 (A)	1979	20700483	20700562	3600	IO-520-F	300*
36-14	TURBO STATIONAIR 7		20700483	20700562	3800	TSIO-520-M	310*
36-04	STATIONAIR 8 (A)	1980	20700563	20700654	3800	IO-520-F	300*
36-14	TURBO STATIONAIR 8		20700563	20700654	3800	TSIO-520-M	310*
36-04	STATIONAIR 8 (A)	1981	20700655	20700729	3800	IO-520-F	300*
36-14	TURBO STATIONAIR 8		20700655	20700729	3800	TSIO-520-M	310*
36-04	STATIONAIR 8 (A)	1982	20700730	20700762	3800	IO-520-F	300*
36-14	TURBO STATIONAIR 8		20700730	20700762	3800	TSIO-520-M	310*
36-04	STATIONAIR 8 (A)	1983	20700763	20700767	3800	IO-520-F	300*
36-14	TURBO STATIONAIR 8		20700763	20700767	3800	TSIO-520-M	310*
36-04	STATIONAIR 8 (A)	1984	20700768	20700788	3800	IO-520-F	300*
36-14	TURBO STATIONAIR 8		20700768	20700788	3800	TSIO-520-M	310*
37-02	208 CARAVAN I	†1985	20800001	20800060	7300	PT6A-114	600
37-03	208A CARAVAN I	†1985	20800001	20800060	8000	PT6A-114	600
37-02	208 CARAVAN I	†1985	20800061	20800076	8000	PT6A-114	600
37-03	208A CARAVAN I	†1985	20800061	20800076	8000	PT6A-114	600
37-02	208 CARAVAN I	†1986	20800077		8000	PT6A-114	600
37-03	208A CARAVAN I	†1986	20800077		8000	PT6A-114	600
34-02	210	1960	57001	57575	2900	IO-470-E	260
34-08	210A	1961	21057576	21057840	2900	IO-470-E	260
34-10	210B	1962	21057841	21058085	3000	IO-470-S	260
34-12	210C	1963	21058086	21058220	3000	IO-470-S	260
34-14	210 CENTURION (D)	1964	21058221	21058510	3100	IO-520-A	285
34-16	210 CENTURION (E)	1965	21058511	21058715	3100	IO-520-A	285
34-18	210 CENTURION (F)	1966	21058716	21058818	3300	IO-520-A	285
34-22	TURBO-SYSTEM CENTURION		T210-0001	T210-0197	3300	TSIO-520-C	285
34-30	210 CENTURION (G)	1967	21058819	21058936	3400	IO-520-A	285
34-32	TURBO-SYSTEM CENTURION		T210-0198	T210-0307	3400	TSIO-520-C	285
34-36	210 CENTURION (H)	1968	21058937	21059061	3400	IO-520-A	285
34-38	TURBO-SYSTEM CENTURION		T210-0308	T210-0392	3400	TSIO-520-C	285
34-39	210 CENTURION (J)	1969	21059062	21059199	3400	IO-520-J	285
34-40	TURBO-SYSTEM CENTURION		T210-0393	T210-0454	3400	TSIO-520-H	285
34-46	CENTURION (K)	1970	21059200	21059351	3800	IO-520-L	300*
34-47	TURBO CENTURION		21059200	21059351	3800	TSIO-520-H	285
34-46	CENTURION (K)	1971	21059352	21059502	3800	IO-520-L	300*
34-47	TURBO CENTURION		21059352	21059502	3800	TSIO-520-H	285
34-48	CENURION (L)	1972	21059503	21059719	3800	IO-520-L	300*
34-49	TURBO CENTURION		21059503	21059719	3800	TSIO-520-H	285
34-48	CENTURION (L)	1973	21059720	21060089	3800	IO-520-L	300*
34-49	TURBO CENTURION		21059720	21060089	3800	TSIO-520-H	285

†INDICATES APPROXIMATE YEAR OF MANUFACTURE ONLY. SIGNIFICANT CHANGES ARE MADE FOR A SERIAL BLOCK, NOT MODEL YEAR.

*5 MIN. TAKEOFF RATING. CONTINUOUS MAXIMUM RATING IS 285 HORSEPOWER

FAA MODEL-SERIAL CODE	MODEL	YEAR	SERIALS Beginning	SERIALS Ending	GROSS/T.O. WT.	ENGINE Model	ENGINE Horsepower
34-48	CENTURION (L)	1974	21060090	21060539	3800	IO-520-L	300*
34-49	TURBO CENTURION		21060090	21060539	3800	TSIO-520-H	285
34-48	CENTURION (L)	1975	21060540	21061039	3800	IO-520-L	300*
34-49	TURBO CENTURION		21060540	21061039	3800	TSIO-520-H	285
34-48	CENTURION (L)	1976	21061040	21061573	3800	IO-520-L	300*
34-49	TURBO CENTURION		21061040	21061573	3800	TSIO-520-H	285
34-50	CENTURION (M)	1977	21061574	21062273	3800	IO-520-L	300*
34-51	TURBO CENTURION		21061574	21062273	3800	TSIO-520-R	310*
34-50	CENTURION (M)	1978	21062274	21062954	3800	IO-520-L	300*
34-51	TURBO CENTURION		21062274	21062954	3800	TSIO-520-R	310*
34-53	CENTURION (N)	1979	21062955	21063640	3800	IO-520-L	300*
34-56	TURBO CENTURION		21062955	21063640	400	TSIO-520-R	310*
34-53	CENTURION (N)	1980	21063641	21064135	3800	IO-520-L	300*
34-56	TURBO CENTURION		21063641	21064135	4000	TSIO-520-R	310*
34-53	CENTURION (N)	1981	21064136	21064535	3800	IO-520-L	300*
34-56	TURBO CENTURION		21064136	21064535	4000	TSIO-520-R	310*
34-53	CENTURION (N)	1982	21064536	21064772	3800	IO-520-L	300*
34-56	TURBO CENTURION		21064536	21064772	4000	TSIO-520-R	310*
34-53	CENTURION (N)	1983	21064773	21064822	3800	IO-520-L	300*
94-56	TURBO CENTURION		21064773	21064822	4000	TSIO-520-R	310*
34-53	CENTURION (N)	1984	21064823	21064897	3800	IO-520-L	300*
34-56	TURBO CENTURION		21064823	21064897	4000	TSIO-520-R	310*
34-58	CENTURION (R)	1985	21064898	21064949	3850	IO-520-L	300*
34-59	TURBO CENTURION		21064898	21064949	4100	TSIO-520-CE	325
34-58	CENTURION (R)	1986	21064950		3850	IO-520-L	300*
34-59	TURBO CENTURION		21064950		4100	TSIO-520-CE	325
34-54	PRESSURIZED CENTURION (N)	1978	P21000001	P21000150	4000	TSIO-520-P	310*
34-54	PRESSURIZED CENTURION (N)	1979	P21000151	P21000385	4000	TSIO-520-P	310*
34-54	PRESSURIZED CENTURION (N)	1980	P21000386	P21000590	4000	TSIO-520-P	310*
34-54	PRESSURIZED CENTURION (N)	1981	P21000591	P21000760	4000	TSIO-520-P	310*
34-54	PRESSURIZED CENTURION (N)	1982	P21000761	P21000811	4000	TSIO-520-AF	310*
34-54	PRESSURIZED CENTURION (N)	1983	P21000812	P21000834	4000	TSIO-520-AF	310*
34-54	PRESSURIZED CENTURION (N)	1984	None	None			
34-55	PRESSURIZED CENTURION (R)	1985	P21000835	P21000866	4100	TSIO-520-CE	325
34-55	PRESSURIZED CENTURION (R)	1986	P21000867		4100	TSIO-520-CE	325
38-20	T303 CRUSADER	1982	T30300001	T30300175	5150	TSIO/LTSIO-520-AE	250
38-20	T303 CRUSADER	1983	T30300176	T30300247	5150	TSIO/LTSIO-520-AE	250
38-20	T303 CRUSADER	1984	T30300258	T30300315	5150	TSIO/LTSIO-520-AE	250
42-02	310	1955	35000	35225	4600	O-470-B	240
42-02	310	1956 & 1957	35226	35546	4600	O-470-B	240
42-08	310B	1957 & 1958	35547	35771	4700	O-470-M	240
42-10	310C	1959	35772	35999	4830	IO-470-D	260
42-10			39001	39031	4830	IO-470-D	260
42-12	310D	1960	39032	39299	4830	IO-470-D	260
42-16	310F	1961	310-0001	310-0156	4830	IO-470-D	260
42-18	310G	1962	310G0001	310G0156	4990	IO-470-D	260
42-20	310H	1963	310H0001	310H0148	5100	IO-470-D	260
42-24	310I	1964	310I0001	310I0200	5100	IO-470-U	260
42-26	310J	1965	310J0001	310J0200	5100	IO-470-U	260
42-28	310K	1966	310K0001	310K0245	5200	IO-470-V	260
42-30	310L	1967	310L0001	310L0207	5200	IO-470-V	260
42-34	310N	1968	310N0001	310N0198	5200	IO-470-V-O	260
42-38	310P	1969	310P0001	310P0240	5200	IO-470-V-O	260
42-40	TURBO-SYSTEM 310P		310P0001	310P0240	5400	TSIO-520-B	285
42-42	310Q	1970	310Q0001	310Q0130	5300	IO-470-V-O	260
42-44	TURBO-SYSTEM 310Q		310Q0001	310Q0130	5500	TSIO-520-B	285
42-42	310Q	1971	310Q0201	310Q0291	5300	IO-470-V-O	260
42-44	TURBO 310Q		310Q0201	310Q0291	5500	TSIO-520-B	285
42-42	310Q	1972	310Q0401	310Q0545	5300	IO-470-V-O	260
42-44	TURBO 310Q		310Q0401	310Q0545	5500	TSIO-520-B	285
42-42	310Q	1973	310Q0601	310Q0845	5300	IO-470-V-O	260
42-44	TURBO 310Q		310Q0601	310Q0845	5500	TSIO-520-B	285
42-42	310Q	1974	310Q0901	310Q1160	5300	IO-470-V-O	260
42-44	TURBO 310Q		310Q0901	310Q1160	5500	TSIO-520-B	285

*5 MIN. TAKEOFF RATING. CONTINUOUS MAXIMUM RATING IS 285 HORSEPOWER.

FAA MODEL-SERIAL CODE	MODEL	YEAR	SERIALS Beginning	SERIALS Ending	GROSS/T.O. WT.	ENGINE Model	ENGINE Horsepower
42-45	310R	1975	310R0001	310R0330	5500	IO-520-M	285
42-46	TURBO 310R		310R0001	310R0330	5500	TSIO-520-B	285
42-45	310R	1976	310R0501	310R0735	5500	IO-520-M	285
42-46	TURBO 310R		310R0501	310R0735	5500	TSIO-520-B	285
42-45	310R	1977	310R0801	310R1004	5500	IO-520-M	285
42-46	TURBO 310R		310R0801	310R1004	5500	TSIO-520-B	285
42-45	310R	1978	310R1201	310R1434	5500	IO-520-M	285
42-46	TURBO 310R		310R1201	310R1434	5500	TSIO-520-B	285
42-45	310R	1979	310R1501	310R1690	5500	IO-520-MB	285
42-46	TURBO 310R		310R1501	310R1690	5500	TSIO-520-BB	285
42-45	310R	1980	310R1801	310R1899	5500	IO-520-MB	285
42-46	TURBO 310R		310R1801	310R1899	5500	TSIO-520-BB	285
42-45	310R	1981	310R2101	310R2140	5500	IO-520-MB	285
42-46	TURBO 310R		310R2101	310R2140	5500	TSIO-520-BB	285
45-02	SKYKNIGHT	1962	320-0001	320-0110	4990	TSIO-470-B	260
45-06	SKYKNIGHT (A)	1963	320A0001	320A0047	5200	TSIO-470-B	260
45-08	SKYKNIGHT (B)	1964	320B0001	320B0062	5200	TSIO-470-C	260
45-10	SKYKNIGHT (C)	1965	320C0001	320C0073	5200	TSIO-470-D	260
45-12	EXECUTIVE SKYKNIGHT (D)	1966	320D0001	320D0130	5200	TSIO-520-B	285
45-14	EXECUTIVE SKYKNIGHT (E)	1967	320E0001	320E0110	5300	TSIO-520-B	285
45-16	EXECUTIVE SKYKNIGHT (F)	1968	320F0001	320F0045	5300	TSIO-520-B	285
56-01	335	1980	335-0001	335-0065	5990	TSIO-520-EB	300
56-02	SKYMASTER	1964	336-0001	336-0195	3900	IO-360-A	210
57-02	SUPER SKYMASTER	1965	33700001	33700239	4200	IO-360-C & D	210
57-04	SUPER SKYMASTER (A)	1966	33700240	33700525	4200	IO-360-C & D	210
57-06	SUPER SKYMASTER (B)	1967	33700526	33700755	4300	IO-360-C & D	210
57-07	TURBO Engine Option		33700526	33700755	4300	TSIO-360 A/B	210
57-12	SUPER SKYMASTER (C)	1968	33700756	33700978	4400	IO-360-C	210
57-14	TURBO Engine Option		33700756	33700978	4500	TSIO-360-A/B	210
57-17	SUPER SKYMASTER (D)	1969	33700979	33701193	4400	IO-360-C	210
57-19	TURBO Engine Option		33700979	33701193	4500	TSIO-360-A	210
57-21	SUPER SKYMASTER (E)	1970	33701194	33701316	4440	IO-360-C	210
57-23	TURBO SUPER SKYMASTER		33701194	33701316	4630	TSIO-360-A	210
57-25	SUPER SKYMASTER (F)	1971	33701317	33701398	4630	IO-360-C	210
57-24	TURBO SUPER SKYMASTER		33701317	33701398	4630	TSIO-360-A	210
57-25	SKYMASTER (F)	1972	33701399	33701462	4630	IO-360-C	210
57-30	SKYMASTER (G)	1973	33701463	33701550	4630	IO-360-G	210
57-30	SKYMASTER (G)	1974	33701551	33701606	4630	IO-360-G	210
57-30	SKYMASTER (G)	1975	33701607	33701671	4630	IO-360-G	210
57-30	SKYMASTER (G)	1976	33701672	33701748	4630	IO-360-G	210
57-30	SKYMASTER (G)	1977	33701749	33701815	4630	IO-360-G	210
57-32	SKYMASTER (H)	1978	33701816	33701874	4630	IO-360-G	210
57-33	TURBO SKYMASTER		33701816	33701874	4630	TSIO-360-H	210
57-32	SKYMASTER (H)	1979	33701875	33701921	4630	IO-360-G	210
57-33	TURBO SKYMASTER		33701875	33701921	4630	TSIO-360-H	210
57-32	SKYMASTER (H)	1980	33701922	33701951	4630	IO-360-GB	210
57-33	TURBO SKYMASTER		33701922	33701951	4630	TSIO-360-HB	210
57-26	PRESSURIZED SKYMASTER (G)	1973	P3370001	P3370148	4700	TSIO-360-C	225
57-26	PRESSURIZED SKYMASTER (G)	1974	P3370149	P3370193	4700	TSIO-360-C	225
57-26	PRESSURIZED SKYMASTER (G)	1975	P3370194	P3370225	4700	TSIO-360-C	225
57-26	PRESSURIZED SKYMASTER (G)	1976	P3370226	P3370257	4700	TSIO-360-C	225
57-26	PRESSURIZED SKYMASTER (G)	1977	P3370258	P3370292	4700	TSIO-360-C	225
57-31	PRESSURIZED SKYMASTER (H)	1978	P3370293	P3370318	4700	TSIO-360-C	225
57-31	PRESSURIZED SKYMASTER (H)	1979	P3370319	P3370341	4700	TSIO-360-C	225
57-31	PRESSURIZED SKYMASTER (H)	1980	P3370342	P3370356	4700	TSIO-360-CB	225
64-04	340	1972	340-0001	340-0115	5975	TSIO-520-K	285
64-04	340	1973	340-0151	340-0260	5975	TSIO-520-K	285
64-04	340	1974	340-0301	340-0370	5975	TSIO-520-K	285
64-04	340	1975	340-0501	340-0555	5975	TSIO-520-K	285
64-05	340A	1976	340A0001	340A0125	5990	TSIO-520-N	310
64-05	340A	1977	340A0201	340A0375	5990	TSIO-520-N	310
64-05	340A	1978	340A0401	340A0562	5990	TSIO-520-N	310
64-05	340A	1979	340A0601	340A0801	5990	TSIO-520-NB	310
64-05	340A	1980	340A0901	340A1045	5990	TSIO-520-NB	310
64-05	340A	1981	340A1201	340A1280	5990	TSIO-520-NB	310
64-05	340A	1982	340A1501	340A1543	5990	TSIO-520-NB	310
64-05	340A	1983	None	None			
64-05	340A	1984	340A1801	340A1817	5990	TSIO-520-NB	310

FAA MODEL-SERIAL CODE	MODEL	YEAR	SERIALS Beginning	Ending	GROSS/T.O. WT.	ENGINE Model	Horsepower
590C/59-0K	401 & 402	1967 & 1968	401/402-0001	401/402-0322	6300	TSIO-520-E	300
59-0D	401A	1969	401A0001	401A0132	6300	TSIO-520-E	300
59-0E	401B	1970 & 1971	401B0001	401B0121	6300	TSIO-520-E	300
59-0E	401B	1972	401B0201	401B0221	6300	TSIO-520-E	300
59-0M	402A	1969	402A0001	402A0129	6300	TSIO-520-E	300
59-0P	402B	1970 & 1971	402B0001	402B0122	6300	TSIO-520-E	300
59-0P	402B	1972	402B0201	402B0249	6300	TSIO-520-E	300
59-0P	402B	1973	402B0301	402B0455	6300	TSIO-520-E	300
59-0P	402B	1974	402B0501	402B0640	6300	TSIO-520-E	300
59-0P	402B	1975	402B0801	402B0935	6300	TSIO-520-E	300
59-0P	402B	1976	402B1001	402B1100	6300	TSIO-520-E	300
59-0P	402B	1977	402B1201	402B1250	6300	TSIO-520-E	300
59-0P	402B	1978	402B1301	402B1384	6300	TSIO-520-E	300
59-0R	402C	1979	402C0001	402C0125	6850	TSIO-520-VB	325
59-0R	402C	1980	402C0201	402C0355	6850	TSIO-520-VB	325
59-0R	402C	1981	402C0401	402C0528	6850	TSIO-520-VB	325
59-0R	402C	1982	402C0601	402C0653	6850	TSIO-520-VB	325
59-0R	402C	1983	None	None			
59-0R	402C	1984	402C0801	402C0807	6850	TSIO-520-VB	325
59-0R	402C	1985	402C0808	402C1022	6850	TSIO-520-VB	325
59-01	404 TITAN	1977	404-0001	404-0136	8400	GTSIO-520-M	375
59-01	404 TITAN	1978	404-0201	404-0246	8400	GTSIO-520-M	375
59-01	404 TITAN	1979	404-0401	404-0460	8400	GTSIO-520-M	375
59-01	404 TITAN	1980	404-0601	404-0695	8400	GTSIO-520-M	375
59-01	404 TITAN	1981	404-0801	404-0859	8400	GTSIO-520-M	375
59-02	411	1965 & 1966	411-0001	411-0250	6500	GTSIO-520-C	340
59-04	411A	1967 & 1968	411A0251	411A0300	6500	GTSIO-520-C	340
59-08	414	1970	414-0001	414-0099	6350	TSIO-520-J	310
59-08	414	1971	414-0151	414-0175	6350	TSIO-520-J	310
59-08	414	1972	414-0251	414-0280	6350	TSIO-520-J	310
59-08	414	1973	414-0351	414-0437	6350	TSIO-520-J	310
59-08	414	1974	414-0451	414-0550	6350	TSIO-520-J	310
59-08	414	1975	414-0601	414-0655	6350	TSIO-520-J	310
59-08	414	1976	414-0801	414-0855	6350	TSIO-520-N	310
59-08	414	1977	414-0901	414-0965	6350	TSIO-520-N	310
59-07	414A CHANCELLOR	1978	414A0001	414A0121	6750	TSIO-520-N	310
59-07	414A CHANCELLOR	1979	414A0201	414A0340	6750	TSIO-520-NB	310
59-07	414A CHANCELLOR	1980	414A0401	414A0535	6750	TSIO-520-NB	310
59-07	414A CHANCELLOR	1981	414A0601	414A0680	6750	TSIO-520-NB	310
59-07	414A CHANCELLOR	1982	414A0801	414A0858	6750	TSIO-520-NB	310
59-07	414A CHANCELLOR	1983	None	None			
59-07	414A CHANCELLOR	1984	414A1001	414A1006	6750	TSIO-520-NB	310
59-07	414A CHANCELLOR	1985	414A1007	414A1212	6750	TSIO-520-NB	310
60-10	421	1967 & 1968	421-0001	421-0200	6800	GTSIO-520-D	375
60-12	421A	1969	421A0001	421A0158	6840	GTSIO-520-D	375
60-14	421B GOLDEN EAGLE & EXECUTIVE COMMUTER	1970	421B0001	421B0056	7250	GTSIO-520-H	375
60-14	421B GOLDEN EAGLE & EXECUTIVE COMMUTER	1971	421B0101	421B0147	7250	GTSIO-520-H	375
60-14	421B GOLDEN EAGLE & EXECUTIVE COMMUTER	1972	421B0201	421B0275	7450	GTSIO-520-H	375
60-14	421B GOLDEN EAGLE & EXECUTIVE COMMUTER	1973	421B0301	421B0486	7450	GTSIO-520-H	375
60-14	421B GOLDEN EAGLE & EXECUTIVE COMMUTER	1974	421B0501	421B0665	7450	GTSIO-520-H	375
60-14	421B GOLDEN EAGLE & EXECUTIVE COMMUTER	1975	421B0801	421B0970	7450	GTSIO-520-H	375
60-16	421C GOLDEN EAGLE & EXECUTIVE COMMUTER	1976	421C0001	421C0171	7450	GTSIO-520-L	375
60-16	421C GOLDEN EAGLE & EXECUTIVE COMMUTER	1977	421C0201	421C0350	7450	GTSIO-520-L	375
60-16	421C GOLDEN EAGLE	1978	421C0401	421C0525	7450	GTSIO-520-L	375
60-16	421C GOLDEN EAGLE	1979	421C0601	421C0715	7450	GTSIO-520-L	375
60-16	421C GOLDEN EAGLE	1980	421C0801	421C0910	7450	GTSIO-520-L	375

FAA MODEL-SERIAL CODE	MODEL	YEAR	SERIALS		GROSS/T.O. WT.	ENGINE	
			Beginning	Ending		Model	Horsepower
60-16	421C GOLDEN EAGLE	1981	421C1001	421C1115	7450	GTSIO-520-N	375
60-16	421C GOLDEN EAGLE	1982	421C1201	421C1257	7450	GTSIO-520-N	375
60-16	421C GOLDEN EAGLE	1983	None	None			
60-16	421C GOLDEN EAGLE	1984	421C1401	421C1413	7450	GTSIO-520-N	375
60-16	421C GOLDEN EAGLE	1985	421C1801	421C1807	7450	GTSIO-520-N	375
60-18	425 CORSAIR	†1980	425-0001	425-0030	8200	PT6A-112	450
60-18	425 CORSAIR	†1981	425-0031	425-0141	8200	PT6A-112	450
60-18	425 CORSAIR	†1982	425-0142	425-0176	8200	PT6A-112	450
60-18	425 CONQUEST I	†1983	425-0177	425-0186	8600	PT6A-112	450
60-18	425 CONQUEST I	†1984	425-0187	425-0221	8600	PT6A-112	450
60-18	425 CONQUEST I	†1985	425-0222	425-0236	8600	PT6A-112	450
60-20	441 CONQUEST II (FORMERLY CONQUEST)	†1977	441-0001	441-0050	9850	TPE331-8	635.5
60-20	441 CONQUEST II (FORMERLY CONQUEST)	†1978	441-0051	441-0096	9850	TPE331-8	635.5
60-20	441 CONQUEST II (FORMERLY CONQUEST)	†1979	441-0097	441-0174	9850	TPE331-8	635.5
60-20	441 CONQUEST II (FORMERLY CONQUEST)	†1980	441-0175	441-0204	9850	TPE331-8	635.5
60-20	441 CONQUEST II (FORMERLY CONQUEST)	†1981	441-0205	441-0277	9850	TPE331-8	635.5
60-20	441 CONQUEST II (FORMERLY CONQUEST)	†1982	441-0278	441-0311	9850	TPE331-8	635.5
60-20	441 CONQUEST II	†1983	441-0312	441-0333	9850	TPE331-8	635.5
60-20	441 CONQUEST II	†1984	441-0334	441-0344	9850	TPE331-8	635.5
60-20	441 CONQUEST II	†1985	441-0345	441-0357	9850	TPE331-8	635.5

†INDICATES APPROXIMATE YEAR OF MANUFACTURE ONLY SIGNIFICANT CHANGES ARE MADE FOR A SERIAL BLOCK, NOT MODEL YEAR.

ARGENTINA & REIMS AVIATION MODELS

CESSNA MODEL-SERIES CODE	MODEL	YEAR	SERIALS Beginning	SERIALS Ending	GROSS/T.O. WT.	ENGINE Model	ENGINE Horsepower
A-01	A-150L	1972	A-1501001	A-1501018	1600	0-200-A	100
A-01	A-150L	1973	A-1501019	A-1501039	1600	0-200-A	100
F-01	F150 (F)	1966	F150-0001	F150-0067	1600	0-200-A	100
F-01	F150 (G)	1967	F150-0068	F150-0219	1600	0-200-A	100
F-01	F150 (H)	1968	F150-0220	F150-0389	1600	0-200-A	100
F-01	F150 (J)	1969	F150-0390	F150-0529	1600	0-200-A	100
F-01	REIMS/CESSNA F150 (K)	1970	F15000530	F15000658	1600	0-200-A	100
F-01	REIMS/CESSNA F150 (L)	1971	F15000659	F15000738	1600	0-200-A	100
F-01	REIMS/CESSNA F150 (L)	1972	F15000739	F15000863	1600	0-200-A	100
F-01	REIMS/CESSNA F150 (L)	1973	F15000864	F15001013	1600	0-200-A	100
F-01	REIMS/CESSNA F150 (L)	1974	F15001014	F15001143	1600	0-200-A	100
F-01	REIMS/CESSNA F150 (M)	1975	F15001144	F15001248	1600	0-200-A	100
F-01	REIMS/CESSNA F150 (M)	1976	F15001249	F15001338	1600	0-200-A	100
F-01	REIMS/CESSNA F150 (M)	1977	F15001339	F15001428	1600	0-200-A	100
A-02	A-A150L AEROBAT	1972	A-A1500001	A-A1500006	1600	0-200-A	100
A-02	A-A150L AEROBAT	1973	A-A1500007	A-A1500009	1600	0-200-A	100
F-03	REIMS/CESSNA AEROBAT (FA150K)	1970	FA1500001	FA1500081	1600	0-200-A	100
F-03	REIMS/CESSNA AEROBAT (FA150L)	1971	FA1500082	FA1500120	1600	0-200-A	100
F-03	REIMS/CESSNA AEROBAT (FRA150L)	1972	FRA1500121	FRA1500166	1650	0-240-A	130
F-03	REIMS/CESSNA AEROBAT (FRA150L)	1973	FRA1500167	FRA1500211	1650	0-240-A	130
F-03	REIMS/CESSNA AEROBAT (FRA150L)	1974	FRA1500212	FRA1500261	1650	0-240-A	130
F-03	REIMS/CESSNA AEROBAT (FRA150M)	1975	FRA1500262	FRA1500281	1650	0-240-A	130
F-03	REIMS/CESSNA AEROBAT (FRA150M)	1976	FRA1500282	FRA1500311	1650	0-240-A	130
F-03	REIMS/CESSNA AEROBAT (FRA150M)	1977	FRA1500312	FRA1500336	1650	0-240-A	130
F-04	REIMS/CESSNA F152	1978	F15201429	F15201528	1670	0-235-L2C	110
F-04	REIMS/CESSNA F152	1979	F15201529	F15201673	1670	0-235-L2C	110
F-04	REIMS/CESSNA F152	1980	F15201674	F15201808	1670	0-235-L2C	110
F-04	REIMS/CESSNA F152	1981	F15201809	F15201893	1670	0-235-L2C	110
F-04	REIMS/CESSNA F152	1982	F15201894	F15201928	1670	0-235-L2C	110
F-04	REIMS/CESSNA F152	1983	F15201929	F15201943	1670	0-235-N2C	108
F-04	REIMS/CESSNA F152	1984	F15201944	F15201952	1670	0-235-N2C	108
F-04	REIMS/CESSNA F152	1985	F15201953	F15201965	1670	0-235-N2C	108
F-04	REIMS/CESSNA F152	1986	F15201966		1670	0-235-N2C	108
F-02	REIMS/CESSNA FA152	1978	FA1520337	FA1520347	1670	0-235-L2C	110
F-02	REIMS/CESSNA FA152	1979	FA1520348	FA1520357	1670	0-235-L2C	110
F-02	REIMS/CESSNA FA152	1980	FA1520358	FA1520372	1670	0-235-L2C	110
F-02	REIMS/CESSNA FA152	1981	FA1520373	FA1520377	1670	0-235-L2C	110
F-02	REIMS/CESSNA FA152	1982	FA1520378	FA1520382	1670	0-235-L2C	110
F-02	REIMS/CESSNA FA152	1983	FA1520383	FA1520387	1670	0-235-N2C	108
F-02	REIMS/CESSNA FA152	1984	None	None			
F-02	REIMS/CESSNA FA152	1985	FA1520388	FA1520415	1670	0-235-N2C	108
F-02	REIMS/CESSNA FA152	1986	FA1520416		1670	0-235-N2C	108
A-03	FP172	1963	FP172-0001	FP172-0003	2500	GO-300-E	175
F-07	F172 (D)	1963	F172-0001	F172-0018	2300	0-300-D	145
F-07	F172 (E)	1964	F172-0019	F172-0085	2300	0-300-D	145
F-07	F172 (F)	1965	F172-0086	F172-0179	2300	0-300-D	145
F-07	F172 (G)	1966	F172-0180	F172-0319	2300	0-300-D	145
F-07	F172 (H)	1967	F172-0320	F172-0431	2300	0-300-D	145
F-07			F172-0436	F172-0442	2300	0-300-D	145
F-07			F172-0444	F172-0446	2300	0-300-D	145
F-07	F172 (H)	1968	F172-0432	F172-0435	2300	0-300-D	145
F-07			F172-0443	F172-0443	2300	0-300-D	145
F-07			F172-0447	F172-0559	2300	0-300-D	145
F-07	F172 (H)	1969	F172-0560	F172-0654	2300	0-300-D	145
F-07	REIMS/CESSNA F172 (H)	1970	F17200655	F17200754	2300	0-300-D	145
F-07	REIMS/CESSNA F172 (K)	1971	F17200755	F17200804	2300	0-300-D	145
F-07	REIMS/CESSNA F172 (L)	1972	F17200805	F17200904	2300	0-320-E2D	150
F-07	REIMS/CESSNA F172 (M)	1973	F17200905	F17201034	2300	0-320-E2D	150
F-07	REIMS/CESSNA F172 (M)	1974	F17201035	F17201234	2300	0-320-E2D	150
F-07	REIMS/CESSNA F172 (M)	1975	F17201235	F17201384	2300	0-320-E2D	150
F-07	REIMS/CESSNA F172 (M)	1976	F17201385	F17201514	2300	0-320-E2D	150
F-07	REIMS/CESSNA F172 (N)	1977	F17201515	F17201639	2300	0-320-H2AD	160
F-07	REIMS/CESSNA F172 (N)	1978	F17201640	F17201749	2300	0-320-H2AD	160
F-07	REIMS/CESSNA F172 (N)	1979	F17201750	F17201909	2300	0-320-H2AD	160
F-07	REIMS/CESSNA F172 (N)	1980	F17201910	F17202039	2300	0-320-H2AD	160
F-07	REIMS/CESSNA F172 (P)	1981	F17202040	F17202134	2400	0-320-D2J	160
F-07	REIMS/CESSNA F172 (P)	1982	F17202135	F17202194	2400	0-320-D2J	160
F-07	REIMS/CESSNA F172 (P)	1983	F17202195	F17202216	2400	0-320-D2J	160

ARGENTINA & REIMS AVIATION MODELS

FAA Model-SERIAL CODE	MODEL	YEAR	SERIALS Beginning	SERIALS Ending	GROSS/T.O. WT.	ENGINE Model	ENGINE Horsepower
F-07	REIMS/CESSNA F172 (P)	1984	F17202217	F17202233	2400	0-320-D2J	160
F-07	REIMS/CESSNA F172 (P)	1985	F17202234	F17202238	2400	0-320-D2J	160
F-07	REIMS/CESSNA F172 (P)	1986	F17202239		2400	0-320-D2J	160
F-08	REIMS ROCKET (FR172E)	1968	FR17200001	FR17200060	2500	IO-360-D	210
F-08	REIMS ROCKET (FR172F)	1969	FR17200061	FR17200145	2500	IO-360-D	210
F-08	REIMS ROCKET (FR172G)	1970	FR17200146	FR17200225	2550	IO-360-D	210
F-08	REIMS ROCKET (FR172H)	1971	FR17200226	FR17200275	2550	IO-360-D	210
F-08	REIMS ROCKET (FR172H)	1972	FR17200276	FR17200350	2550	IO-360-D	210
F-08	REIMS ROCKET (FR172J)	1973	FR17200351	FR17200440	2550	IO-360-H	210
F-08	REIMS ROCKET (FR172J)	1974	FR17200441	FR17200530	2550	IO-360-H	210
F-08	REIMS ROCKET (FR172J)	1975	FR17200531	FR17200559	2550	IO-360-J	210
F-08	REIMS ROCKET (FR172J)	1976	FR17200562	FR17200590	2550	IO-360-J	210
F-08	REIMS/CESSNA HAWK XP (FR172K)	1977	FR17200591	FR17200620	2550	IO-360-K	195
F-08	REIMS/CESSNA HAWK XP (FR172K)	1978	FR17200621	FR17200630	2550	IO-360-K	195
F-08	REIMS/CESSNA HAWK XP (FR172K)	1979	FR17200631	FR17200655	2550	IO-360-K	195
F-08	REIMS/CESSNA HAWK XP (FR172K)	1980	FR17200656	FR17200665	2550	IO-360-K	195
F-08	REIMS/CESSNA HAWK XP (FR172K)	1981	FR17200666	FR17200675	2550	IO-360-K	195
F-09	REIMS/CESSNA CARDINAL RG (F177RG)	1971	F177RG0001	F177RG0042	2800	IO-360-A1B6	200
F-09	REIMS/CESSNA CARDINAL RG (F177RG)	1972	F177RG0043	F177RG0062	2800	IO-360-A1B6	200
F-09	REIMS/CESSNA CARDINAL RG (F177RG)	1973	F177RG0063	F177RG0092	2800	IO-360-A1B6D	200
F-09	REIMS/CESSNA CARDINAL RG (F177RG)	1974	F177RG0093	F177RG0122	2800	IO-360-A1B6D	200
F-09	REIMS/CESSNA CARDINAL RG (F177RG)	1975	F177RG0123	F177RG0138	2800	IO-360-A1B6D	200
F-09	REIMS/CESSNA CARDINAL RG (F177RG)	1976	F177RG0139	F177RG0160	2800	IO-360-A1B6D	200
F-09	REIMS/CESSNA CARDINAL RG (F177RG)	1977	F177RG0161	F177RG0177	2800	IO-360-A1B6D	200
A-12	A182J	1966	A182-0001	A182-0056	2800	0-470-R	230
A-12	A182K	1967	A182-0057	A182-0096	2800	0-470-R	230
A-12	A182L	1968	A182-0097	A182-0116	2800	0-470-R	230
A-12	A182M & N	1969-1971	None	None			
A-12	A182N	1972	A182-0117	A182-0136	2800	0-470-R	230
A-12	A182N	1973	None	None			
A-12	A182N	1974	A182-0137	A182-0146	2950	0-470-R	230
A-12	A182N	1975	None	None			
A-12	A182N	1976	A182-0147	A182-0148	2950	0-470-R	230
F-11	REIMS/CESSNA F182 SKYLANE (P)	1976	F18200001	F18200025	2950	0-470-S	230
F-11	REIMS/CESSNA F182 SKYLANE (Q)	1977	F18200026	F18200064	2950	0-470-U	230
F-11	REIMS/CESSNA F182 SKYLANE (Q)	1978	F18200065	F18200094	2950	0-470-U	230
F-11	REIMS/CESSNA F182 SKYLANE (Q)	1979	F18200095	F18200129	2950	0-470-U	230
F-11	REIMS/CESSNA F182 SKYLANE (Q)	1980	F18200130	F18200169	2950	0-470-U	230
F-12	REIMS/CESSNA SKYLANE RG	1978	FR18200001	FR18200020	3100	0-540-J3C5D	235
F-12	REIMS/CESSNA SKYLANE RG	1979	FR18200021	FR18200045	3100	0-540-J3C5D	235
F-12	REIMS/CESSNA SKYLANE RG	1980	FR18200046	FR18200070	3100	0-540-J3C5D	235
A-13	A-A188B AG TRUCK	1972	A-A1880001	A-A1880008	4000	IO-520-D	300*
A-13	A-A188B AG TRUCK	1973	A-A1880009	A-A1880018	4000	IO-520-D	300*
A-13	A-A188B AG TRUCK	1974	A-A1880019	A-A1880024	4000	IO-520-D	300*
A-13	A-A188B AG TRUCK	1975	None	None			
A-13	A-A188B AG TRUCK	1976	A-A1880025	A-A1880034	4000	IO-520-D	300*
A-14	REIMS SUPER SKYMASTER (E)	1970	F33700001	F33700024	4440	IO-360-C	210
A-14	REIMS TURBO SUPER SKYMASTER (E)	1970	F33700001	F33700024	4630	TSIO-360-A	210
A-14	REIMS SUPER SKYMASTER (F)	1971	F33700025	F33700045	4630	IO-360-C	210
A-14	REIMS TURBO SUPER SKYMASTER (F)	1971	F33700025	F33700045	4630	TSIO-360-A	210
A-14	REIMS/CESSNA SKYMASTER (F)	1972	F33700046	F33700055	4630	IO-360-C	210
A-14	REIMS/CESSNA SKYMASTER (G)	1973	F33700056	F33700063	4630	IO-360-G	210
A-14	REIMS/CESSNA SKYMASTER (G)	1974	F33700064	F33700071	4630	IO-360-G	210
A-14	REIMS/CESSNA SKYMASTER (G)	1975	F33700072	F33700076	4630	IO-360-G	210
A-14	REIMS/CESSNA SKYMASTER (G)	1976	F33700077	F33700079	4630	IO-360-G	210
A-14	REIMS/CESSNA SKYMASTER (G)	1977	F33700080	F33700084	4630	IO-360-G	210
A-14	REIMS/CESSNA SKYMASTER (H)	1978	F33700085	F33700086	4630	IO-360-G	210
A-15	REIMS/CESSNA PRESS. SKYMASTER	1973	FP33700001	FP33700008	4700	TSIO-360-C	225
A-15	REIMS/CESSNA PRESS. SKYMASTER	1974	FP33700009	FP33700013	4700	TSIO-360-C	225
A-15	REIMS/CESSNA PRESS. SKYMASTER	1975	FP33700014	FP33700015	4700	TSIO-360-C	225
A-15	REIMS/CESSNA PRESS. SKYMASTER	1976	FP33700016	FP33700017	4700	TSIO-360-C	225
A-15	REIMS/CESSNA PRESS. SKYMASTER	1977	FP33700018	FP33700022	4700	TSIO-360-C	225
A-15	REIMS/CESSNA PRESS. SKYMASTER	1978	FP33700023	FP33700023	4700	TSIO-360-C	225
	REIMS/CESSNA F406 CARAVAN II	†1985	F406-0001	F406-0006	9360	PT6A-112	500
	REIMS/CESSNA F406 CARAVAN II	†1986	F406-0007		9360	PT6A-112	500

CESSNA AIRCRAFT SERIAL NUMBERS
MILITARY TO COMMERCIAL MODEL CROSS REFERENCE

FAA MODEL-SERIES	MODEL	MIL. DESIG.	YEAR	CESSNA SERIALS Beginning	CESSNA SERIALS Ending	Note	MILITARY SERIALS Beginning	MILITARY SERIALS Ending	
24-14	172F	T-41A	1965	17251947	17253392	1	65-5100	65-5269	USAF
24-20	172G	T-41A	1966	17254405	17254544	1	-	-	Peru
24-24	172H	T-41A	1967	17256230	17256346	1	67-14959	67-14992	USAF
24-30	172K	T-41A	1969	17259003	17259015	1	69-7743	69-7749	USAF
24-13	R172E	T-41B	1967	R172-0001	R172-0255		67-15000	67-15254	US ARMY
24-13	R172E	-	1967	R172-0256	-		-	REIMS PROTO FR1720001	
24-13	R172E	T-41C	1968	R172-0257	R172-0301		68-7866	68-7910	USAF
24-13	R172E	T-41D	1968	R172-0302	R172-0315		68-8944	68-8957	USAF/MAP
24-13	R172E	T-41D	1968	R172-0316	R172-0335	1969	68-8958	68-8977	USAF/MAP
	R172F	T-41D	1969	R172-0336	R172-0365		-	-	COLUMBIA
	R172F	T-41D	1969	R172-0366	R172-0385		69-7181	69-7200	USAF/MAP
	R172F	T-41D	1969	R172-0386	R172-0391		69-7274	69-7279	USAF/MAP
	R172F	T-41D	1969	R172-0392	R172-0409	1969	69-7675	69-7692	USAF/MAP
24-21	R172G	T-41D	1970	R1720410	R1720425	2	70-1592	70-1607	USAF/MAP
24-21	R172G	T-41C	1969	R1720426	R1720432		69-7750	69-7756	USAF
24-21	R172G	T-41D	1970	R1720433	R1720444		-	-	ECUADOR
24-25	R172H	T-41D	1970	R1720445	-		70-1962	-	USAF/MAP
24-25	R172H	T-41D	1970	R1720446	R1720452		70-2021	70-2027	USAF/MAP
24-25	R172H	T-41D	1971	R1720453	R1720454		71-1051	71-1052	USAF/MAP
24-25	R172H	T-41D	1970	R1720455	R1720479		70-2037	70-2061	USAF/MAP
24-25	R172H	T-41D	1971	R1720480	R1720481		71-1053	71-1054	USAF/MAP
24-25	R172H	T-41D	1971	R1720482	R1720485		71-20940	71-20943	USAF/MAP
24-25	R172H	T-41D	1971	R1720486	R1720493		71-1055	71-1062	USAF/MAP
24-25	R172H	T-41D	1970	R1720494	-		70-2456	-	USAF/MAP
24-25	R172H	T-41D	1972	R1720495	R1720496		72-1384	72-1385	USAF/MAP
24-25	R172H	T-41D	1971	R1720497	R1720506		71-1458	71-1467	USAF/MAP
24-25	R172H	T-41D	1971	R1720507	R1720509		71-1334	72-1336	USAF/MAP
24-25	R172H	T-41D	1971	R1720510	R1720539		71-1408	72-1437	USAF/MAP
24-25	R172H	T-41D	1972	R1720540	R1720545		72-1470	72-1475	USAF/MAP
24-25	R172H	T-41D	1972	R1720546	-		-	-	ECUADOR
24-25	R172H	T-41D	1973	R1720547	R1720551		73-1659	73-1663	USAF/MAP
24-25	R172H	T-41D	1973	R1720552	R1720553		-	-	ISRAEL
24-25	R172H	T-41D	1973	R1720554	R1720558		-	-	HONDURAS
24-29	R172J	-	1973	R1720559	-		PROTOTYPE R172J BECAME S/N 680		
24-25	R172H	T-41D	1973	R1720560	-		-	-	ECUADOR
24-25	R172H	T-41D	1973	R1720561	R1720563		-	-	ECUADOR
24-25	R172H	T-41D	1973	R1720564	R1720603		-	-	PERU
24-25	R172H	T-41D	1974	R1720604	-		74-01724	-	USAF/MAP
24-25	R172H	T-41D	1974	R1720605	R1720608		74-02093	74-02096	USAF/MAP
24-25	R172H	T-41D	1975	R1720609	R1720611		75-0732	75-0734	USAF/MAP
24-25	R172H	T-41D	1974	R1720612			74-02113		USAF/MAP
24-25	R172H	T-41D		R1720613	R1720617		AE015	AE055	ARGENTINA
24-25	R172H	T-41D		R1720621	R1720625		-	-	PHILIPPINES
42-04	310A	L-27A (U-3A)	1957	38001	38080		57-5846	57-5925	
42-04	310A	L-27A (U-3A)	1958	38081	38160		58-2107	58-2186	
42-14	310E	U-3B	1960	310M0001	310M0036	3	60-6047	60-6081	
57-08	M337B	O-2A	1967	M337-0001	M337-0467		67-21295	67-21439	USAF
57-04	337A	O-2B	1967	MC337-0240	MC337-0523		67-21440	67-21470	USAF

NOTE: 1 T-41 SERIALS SCATTERED THROUGHOUT THIS S/N BLOCK.

2 IN 1970, THE DASH WAS ELIMINATED, BUT **NOT** REPLACED WITH A "0" ON THE R172, AS IT WAS ON THE OTHER MODELS.

3 THERE WERE 36 310E BUILT: 1 COMMERCIAL DEMO (310M0001) AND 35 U-3B (310M0002/310M0036).

June 8, 1988

Dear Piper Customer:

As the owner of Piper Aircraft Corporation, I am taking this time to urge you to read carefully the attached Service Bulletin to insure your better understanding of its contents and to provide assurance of Piper's commitment to you and your flying safety.

Last year, the Federal Aviation Administration issued Airworthiness Directive 87-08-08 requiring complicated wing spar inspections on certain PA-28 and PA-32 models with more than 5,000 hours flight time.

Following the inspection of more than 500 airplanes in 1987, only two aircraft were found with cracks. (Both PA-32 aircraft had previously suffered extensive airframe damage and continued to be operated in unusually severe environments.)

Piper has conducted extensive metallurgical examinations, investigated several methods of inspection and conducted detailed fatigue and fracture analyses. As a result of these efforts and the 500+ field inspections, the Federal Aviation Administration suspended AD 87-08-08.

The Service Bulletin has been prepared based on the extensive investigations carried out over the past year. It provides a definition of inspections required to assure continued structural integrity of the affected models. The vast majority of owners will find, upon careful examination and study of this Service Bulletin, that the time for initial spar inspection may be years or decades away. However, some aircraft will require spar inspection in the immediate future.

The Piper Service Bulletin No. 886 details inspection parameters and procedures. PLEASE READ THE BULLETIN CAREFULLY AND COMPLETELY. If you have any questions or comments, please call us at 1-800-72-PIPER (1-800-727-4737).

Remember, our aim is to consistently provide you with the best in personal aircraft and support services.

Sincerely,

PIPER AIRCRAFT CORPORATION

M. Stuart Millar

Piper Aircraft Corporation
Vero Beach, Florida, U.S.A.

SERVICE BULLETIN No. 886

PIPER CONSIDERS COMPLIANCE MANDATORY

Date June 8, 1988 S

SUBJECT: Wing Spar Inspection

MODELS AFFECTED:	SERIAL NUMBERS AFFECTED:
GROUP I:	
PA-28-140 Cherokee	28-20000 through 28-7725290
PA-28-150/160/180 Cherokee	28-1 through 28-7505259 and 28-E13
PA-28-151 Warrior	28-7415001 through 28-7715314
PA-28-161 Warrior II	28-7716001 through 28-8616057, 2816001 and up
PA-28-181 Archer II	28-7690001 through 28-8690062, 2890001 and up
PA-28R-180 Arrow	28R-30001 through 28R-7130013
PA-28R-200 Arrow/Arrow II	28R-35001 through 28R-7635545
GROUP II:	
PA-28R-201 Arrow III	28R-7737001 through 28R-7837319
PA-28R-201T Turbo Arrow III	28R-7703001 through 28R-7803374
PA-28RT-201 Arrow IV	28R-7918001 through 28R-8218026
PA-28RT-201T Turbo Arrow IV	28R-7931001 through 28R-8631005, 2831001 and up
PA-28-235 Cherokee	28-10001 through 28-7710089 and 28-E11
PA-32-260 Cherokee Six	32-1 through 32-7800008
PA-32-300 Cherokee Six 300	32-40000 through 32-7940290

COMPLIANCE TIME: (ALL AIRCRAFT)

1. Within the next fifty (50) hours time in service, accomplish INSTRUCTION 1 and INSTRUCTION 2 below.

2. Within 50 hours time in service of the initial and repetitive compliance times determined in instruction 2 below, accomplish INSTRUCTION 3.

PURPOSE:

On March 30, 1987, a PA-28 engaged in pipeline patrol operations suffered an in-flight wing separation resulting in a fatal accident. Investigation revealed the wing failure was due to propagation of a fatigue crack, which originated in the wing lower main spar cap.

F.A.A. issued Airworthiness Directive 87-08-08 requiring wing removal and inspection on many PA-28 and PA-32 series airplanes with more than 5000 hours total time in service. To date, over five-hundred (500) inspections have been accomplished. Only two (2) negative findings were reported on a pair of PA-32's operating in a severe environment and with considerable damage histories.

Based on these 500+ inspections, and extensive wing fatigue and fracture analyses begun by Piper, F.A.A. suspended AD 87-08-08 on September 28, 1987; and published additional information in the General Aviation Airworthiness Alerts (Special Issue Advisory Circular AC 43.16). The Piper fatigue/fracture analysis program is complete, resulting in the inspection requirements contained in this Service Bulletin.

Piper understands that the majority of aircraft are, have been, and will continue to be operated well within the aircraft's design parameters during all of their operational life. It has been determined that aircraft that remain in this "normal usage class" may safely continue to operate tens of thousands of hours before wing removal and inspection is required.

HOWEVER, Piper also realizes that some small number of aircraft engage in operations which, for the purposes of this Service Bulletin, are considered "severe" or "extreme" and require greatly reduced wing removal and inspection intervals.

This Service Bulletin provides instructions for: (1) determining the aircraft's "usage class"; (2) determining the initial and recurring inspection times; and (3) accomplishing the wing spar inspection(s).

The contents of this Service Bulletin are somewhat complicated. Owner/operators should read it CAREFULLY, and conduct a thorough review of their aircraft's operating records to establish the correct compliance times. FAILURE TO FULLY COMPLY WITH THIS SERVICE BULLETIN COULD SERIOUSLY AFFECT THE STRUCTURAL INTEGRITY, SAFETY AND AIRWORTHINESS OF THE AIRCRAFT!

APPROVAL: The technical contents of this Service Bulletin have been approved by the FAA.

INSTRUCTIONS:

Instruction 1, Determination of Aircraft "Usage Class". (See TABLE 1)

NOTE: It is necessary to have complete documentation and/or knowledge of the aircraft's entire operating history, in order to make a valid determination of "Usage Class" and Compliance times.

A. Normal Usage, Class 'A'.

This class applies to all aircraft which do not and have not engaged in operations considered as "Severe", "Extreme", or "Unknown" in the Usage Class described below.

Most aircraft affected by this Service Bulletin will fall into this "Normal Usage Class". Normal flight training operations fall into this class as well. However, if there is any doubt as to the aircraft's operating history, it is recommended that the initial inspection be conducted in accordance with the UNKNOWN USAGE CLASS 'D' Compliance Time.

B. Severe Usage, Class 'B'.

This class applies to aircraft which have engaged in severe usage, involving contour or terrain following operations, (such as power/pipeline patrol, fish/game spotting, aerial application, aerial advertising, police patrol, livestock management or other activities) where a significant part of the total flight time has been spent below one-thousand (1000) feet AGL, altitude.

NOTE: Aircraft with part of total time in service in SEVERE USAGE CLASS 'B' operations and part in NORMAL USAGE CLASS 'A', may adjust compliance times by a "Factored Service Hours" calculation. See Instruction 2A to calculate "Factored Service Hours".

C. Extreme Usage, Class 'C'.

This class applies to aircraft which have been damaged due to operations from extremely rough runways, flight in extreme damaging turbulence or other accident/incident which required major repair or replacement of wing(s), landing gear or engine mount.

D. Unknown Usage, Class 'D'.

This class applies to aircraft and/or wings of unknown or undetermined operational or maintenance history.

Instruction 2, Determination of Initial and Repetitive Compliance Times.

Upon determining aircraft model/serial number "affected" group, from Page 1, and Aircraft "Usage Class" from Instruction 1, determine the applicable initial or repetitive wing spar inspection compliance time from TABLE 1 on Page 7.

Instruction 2A, Factored Service Time Formula.

NOTE: This formula applies only to aircraft in SEVERE USAGE CLASS 'B'. It may be used to calculate the initial and repetitive inspection times in factored hours to afford use of TABLE 1 in Instruction 2 above, provided a portion of their operating time in service has been in "Normal Usage, Class A".

FACTORED SERVICE hours shall be determined as follows:

EXAMPLES (For Group I aircraft)

$$\text{Hours in Severe Service} + \frac{\text{Hours in Normal Service}}{17} = \text{FACTORED SERVICE hours}$$

(1) $2000 + \frac{10000}{17} =$ 2588 Factored Service Hours

(2) $3500 + \frac{8500}{17} =$ 4000 Factored Service Hours

EXAMPLE NO. 1 - (12,000 hours total time in service.)

Initial inspection not required at this time. Will require initial inspection when Total Factored Service Hours reach 3700.

EXAMPLE NO. 2 - (12,000 hours total time in service.)

Initial inspection required within the next 50 hours time in service.

Instruction 3, Wing Spar Inspection Instructions.

To prevent the propagation of cracks in the wing lower spar cap and subsequent separation of the wing, accomplish the following.

(A) Remove both wings in accordance with Piper Service Letter No. 997, dated May 14, 1987.
Caution: Use extreme care in removing and replacing the wing main spar to the fuselage attachment bolts (18 per side) to preclude damaging the bolt holes. Do not drive the bolts in or out of the holes. As the bolts are removed, number each bolt and hole to ensure replacement in the same hole. Use proper torque values when installing bolts. If replacement of some bolts is required, ensure proper part number and grip length. Installation of new MS20365-624C nuts on the main spar attach bolts during wing reinstallation is recommended.

(B) Visually inspect, using a 10-power (minimum) magnifying glass and a dye-penetrant method or equivalent, for cracks in the wing lower spar cap from the wing skin line outboard of the outboard row of wing attach bolt holes to an area midway between the second and third row of bolt holes from the outboard row.

(1) If no cracks are found, prior to further flight, accomplish the actions specified in Paragraph (C) below.
(2) If any cracks are found, prior to further flight, replace the spar or wing with a serviceable unit shown to be free of cracks when subjected to the inspections specified in this paragraph.

(C) Visually inspect for cracks in each upper wing skin adjacent to the fuselage and forward of each main spar.

(1) If no cracks are found, reinstall the wings in accordance with the instructions in the applicable Piper Maintenance Manual for that airplane.
(2) If cracks are found, prior to further flight, repair in accordance with AC 43.13-1, and reinstall the wings in accordance with the instructions in the applicable Piper Maintenance Manual for that airplane.

(D) Make an appropriate logbook entry of compliance with this Service Bulletin.

MATERIAL REQUIRED: To be determined by inspection. Refer to the appropriate aircraft Parts Catalog.

AVAILABILITY OF PARTS: Your Piper Field Service Facility.

EFFECTIVITY DATE: This Service Bulletin is effective upon receipt.

SUMMARY: Please contact your Factory Authorized Piper Field Service Facility to make arrangements for compliance with this Service Bulletin, in accordance with the Compliance Time indicated.

NOTE: If you are no longer in possession of this aircraft, please forward this information to the present owner/operator and notify the Factory of address/ownership corrections. Changes should include aircraft model, serial number, current owner's name and address.

Corrections/Changes should be directed to:

Piper Aircraft Corporation
Attn: Customer Services
P.O. Box 1328
Vero Beach, FL 32961-1328

 COMPLIANCE TIME CHART (TABLE 1)

USAGE CLASS	GROUP I AIRCRAFT		GROUP II AIRCRAFT	
	INITIAL INSPECTION	REPETITIVE INSPECTION	INITIAL INSPECTION	REPETITIVE INSPECTION
NORMAL USAGE CLASS 'A'*	62,900 Hours total time in service*	Every 6000 hours time in service after 62,900 hours total time in service	30,600 Hours total time in service*	Every 3000 hours time in service after 30,600 hours total time in service
SEVERE USAGE CLASS 'B' **(see instruction) 2A	Within 50 hours time in service upon reaching 3,700 hours** total time in service	Every 1,600 hours time in service after initial inspection	Within 50 hours time in service upon reaching 1,800 hours** total time in service	Every 800 hours time in service after initial inspection
EXTREME USAGE CLASS 'C' (Damage History)	Within 50 hours time in service***	Every 1,600 hours time in service after initial inspection	Within 50 hours time in service***	Every 800 hours time in service after initial inspection
UNKNOWN USAGE CLASS 'D' (Aircraft/Wings with unknown service history)	Within 50 hours time in service***	Determined by applicable usage class after initial inspection	Within 50 hours time in service***	Determined by applicable usage class after initial inspection

*NOTE: Aircraft which have complied with AD 87-08-08 in Normal Usage Class 'A', must comply with the initial inspection requirements upon reaching 62,900 hours total time in service for Group I and 30,600 total time in service for Group II.

**NOTE: Aircraft which have previously complied with AD 87-08-08 in Severe Usage Class 'B' are in compliance with the initial inspection requirement, only if the AD 87-08-08 inspection was performed at or after 3,700 hours time in service for Group I or 1,800 hours time in service for Group II.

***NOTE: Aircraft which have previously complied with AD 87-08-08 in CLASS 'C', and 'D' are in compliance with the initial inspection requirements of this Service Bulletin.